Robotics and Automation in Industry 4.0

Edited by

Santosh Kumar Das
Department of Computer Science and Engineering at Sarala Birla University, India

Soumi Majumder
Department of Business Administration, Future Institute of Engineering and Management, India

&

Nilanjan Dey
Department of Computer Science and Engineering at Techno International New Town, India

Robotics and Automation in Industry 4.0

Editors: Santosh Das, Soumi Majumder, Nilanjan Dey

ISBN (Online): 978-981-5223-49-1

ISBN (Print): 978-981-5223-50-7

ISBN (Paperback): 978-981-5223-51-4

First published in 2024.

Bentham Science Publishers Pte. Ltd.
80 Robinson Road #02-00
Singapore 068898
Singapore
Email: subscriptions@benthamscience.net

BENTHAM SCIENCE

CONTENTS

PREFACE i
THE OBJECTIVE OF THE BOOK i
ORGANIZATION OF THE BOOK i
Section 1: Robotics: Applications and Advancements (Chapters 1-3) i
Chapter 1 i
Chapter 2 ii
Chapter 3 ii
Section 2: Application of Renewable Energy: Power for a Sustainable Future (Chapters 4-5) ii
Chapter 4 ii
Chapter 5 ii
Section 3: FinTech: Systematic Analysis and its Applications (Chapters 6-8) ii
Chapter 6 ii
Chapter 7 ii
Chapter 8 ii
Section 4: Multidisciplinary Approach: Understanding, Benefits and Applications (Chapters 9-13) iii
Chapter 9 iii
Chapter 10 iii
Chapter 11 iii
Chapter 12 iii
Chapter 13 iii

LIST OF CONTRIBUTORS iv

CHAPTER 1 NANOROBOTS: TYPES, PRINCIPLES AND APPLICATIONS 1
Parkarsh Kumar and *Santosh Kumar Das*
INTRODUCTION 1
TYPES 3
Applications and Recent Developments 5
CONCLUSION 6
REFERENCES 6

CHAPTER 2 ROBOT PATH PLANNING IN A DYNAMIC ENVIRONMENT USING DEEP Q-LEARNING 9
Rifaqat Ali and *Preeti Chandrakar*
INTRODUCTION 9
Structure of the Paper 11
RELATED WORKS 11
Graph-Based Approach 11
Fuzzy Logic Approach 12
Artificial Neural Network-Based Approach 13
PROPOSED METHODOLOGY 14
The Agent 14
The Environment 15
Reinforcement Learning 15
Markov Decision Process 16
Q-Learning 17
Deep Q-Learning 18
Experience Replay 19
SIMULATION EXPERIMENTS 19
The Simulator 19

Deep Q-Learning ... 21
Basic Workflow ... 23
Definition of Initial and Goal State ... 23
Definition of the Obstacles ... 24
Training and Optimization ... 25
Optimization of the Path ... 26
PERFORMANCE EVALUATION ... 26
CONCLUSION ... 30
REFERENCES ... 30

CHAPTER 3 SEVEN CONCERNS OF INNOVATION DESIGN PROCESS AND ANALYSIS OF A SMART BICYCLE FRAME ... 34
Suraj, Arun Kumar and *Tameshwer Nath*
INTRODUCTION ... 34
Concerns of Innovation ... 35
DESIGN CONCEPTS ... 36
In Design Concept 1 ... 36
Design Concept 2 ... 37
Design Concept 3 ... 38
Design Concept 4 ... 38
ANALYSIS OF FINAL DESIGNED FRAME ... 39
RESULTS AND DISCUSSION ... 40
CONCLUSION ... 41
REFERENCES ... 42

CHAPTER 4 HYBRID OPTIMIZATION OF PROFIT-BASED UNIT COMMITMENT ALLOWING FOR UNCERTAINTIES OF RENEWABLE ENERGY SOURCES IN SUMMER AND WINTERTIME ... 43
Ranadip Roy, Ayani Nandi and *Nirmalya Mallick*
INTRODUCTION ... 44
LITERATURE ANALYSIS ... 44
KEY FEATURES ... 48
METHODOLOGIES AND FORMULATION ... 48
HYBRID VERSION OF THE CHAOTIC APPROACH OF HARRIS HAWKS COMBINED WITH THE SINE COSINE OPTIMIZER ... 50
HYBRID CHAOTIC SLIME MOULD-SINE COSINE OPTIMIZER ... 52
SIMULATION ANALYSIS ... 52
CONCLUSION ... 59
REFERENCES ... 60

CHAPTER 5 THE IMPACT OF INDUSTRY 4.0 ON MANUFACTURING: CHALLENGES AND OPPORTUNITIES ... 64
Ramesh Chandra Goswami, Hiren Joshi and *Sunil Gautam*
INTRODUCTION ... 64
History of the Industrial Revolution ... 65
Industrial Revolution-First ... 66
Industrial Revolution-Second ... 66
Industrial Revolution-Third ... 66
Industrial Revolution-Fourth ... 66
DESIGN PRINCIPLES OF INDUSTRY 4.0 ... 67
TECHNOLOGIES IMPACTING PRODUCTION ... 68
THE PILLARS OF INDUSTRY 4.0 ... 68

Internet of Things (IoT) ... 68
Cloud Computing ... 69
AI and Machine Learning ... 69
Edge Computing ... 70
Cyber Security ... 70
Digital Twin ... 70
Cyber-Physical Systems ... 70
DIGITAL TWIN APPLICATIONS FOR MANUFACTURING ... 70
Cyber-Physical Systems ... 72
THE BASIC CHARACTERISTICS OF INDUSTRY 4.0 [51] ... 74
FEATURES OF A SMART FACTORY ... 75
Networks and Technologies ... 75
Decision Making Process ... 75
Supply Chain ... 75
Production ... 76
CHALLENGES AND OPPORTUNITIES IN THE IMPLEMENTATION OF INDUSTRY 4.0 ... 76
CONCLUSION AND FUTURE SCOPE ... 77
REFERENCES ... 78

CHAPTER 6 EMERGING TECHNOLOGIES IN FINTECH: A CASE STUDY ... 83
Keyurkumar Patel, Pujita Sunnapu and *Sunil Gautam*
INTRODUCTION ... 83
Motivation ... 85
Advantages: ... 85
Limitations: ... 86
LITERATURE REVIEW ... 86
FINTECH ECOSYSTEM ... 87
FINTECH BUSINESS CHALLENGES ... 88
FinTech Business Areas: ... 88
FinTech Business Challenges: ... 89
USE CASE FOR FINTECH TECHNOLOGIES ... 89
Customer Support ... 92
Risk Management ... 93
Robotics Process Automation (RPA) ... 93
Blockchain ... 94
Application of Blockchain in FinTech: ... 94
CONCLUDING REMARKS ... 95
REFERENCES ... 96

CHAPTER 7 LEVERAGING BLOCKCHAIN TECHNOLOGIES IN HEALTHCARE APPLICATIONS ... 98
Manorama Patnaik
INTRODUCTION ... 98
TYPES OF BLOCKCHAIN ... 100
Public Blockchains ... 100
Private Blockchains ... 100
Consortium/Federated Blockchains ... 101
Hybrid Blockchains ... 101
BLOCKCHAIN-BASED IOT HEALTHCARE APPLICATIONS ... 102
Electronic Health Records (EHRs) ... 103
Data Sharing ... 104

Data Storage 105
Data Management 105
CASE STUDY: ETHEREUM-BASED SMART CONTRACT SYSTEM 106
DRAWBACKS OF BLOCKCHAIN IN THE INTERNET OF MEDICAL THINGS 108
CONCLUSION 109
FUTURE SCOPE 109
REFERENCES 110

CHAPTER 8 DESIGN AND DEVELOPMENT OF BLOCKCHAIN INTEGRATED IOT SYSTEM FOR PHARMACEUTICAL APPLICATIONS 114
Jyothy S. T. and *Mrinal Sarvagya*
INTRODUCTION 115
Internet of Things (IoT) 115
Blockchain 117
Components of a Blockchain Architecture 118
LITERATURE SURVEY 120
IOT ENABLED TECHNOLOGIES 128
Wireless Sensor Networks (WSN) 128
Sensors 128
Actuators 128
Cloud Computing 128
Big Data Analysis 128
Communication Protocols 128
Embedded Systems 129
Radio-Frequency Identification 129
CHALLENGES FACED BY TRADITIONAL SUPPLY CHAIN 129
Transparent Supply Chain 130
Temperature Control 130
Warehouse Management 130
Lack of Standardization 130
SOLUTIONS PROVIDED BY BLOCKCHAIN-INTEGRATED INTERNET OF THINGS SUPPLY CHAIN SYSTEM 130
Transparent Supply Chain 131
Temperature Control 131
Smart Warehouse 132
Standardization 132
Telemedicine 132
CONCLUSION 133
REFERENCES 133

CHAPTER 9 ATTRITION IN IT SECTOR: PSYCHOLOGY BEHIND THE SCENE 136
Abhisek Sarkar
INTRODUCTION 136
Factors Leading to Employee Attrition 138
The Science Behind the Motivation 138
Relation Between Attrition and Dopamine 139
Attrition in the IT Industry 140
Literature Review 142
Action of Dopamine and Serotonin in Controlling Emotions 144
Dopamine, Serotonin and Sleep 146
Dopamine, Serotonin and Digestion 146
Dopamine, Serotonin and Mental Health 146

Dopamine 147
Serotonin 147
Attrition Rate in Indian and India-based IT/ITeS firms 147
CONCLUSION 149
REFERENCES 149

CHAPTER 10 DEEP LEARNING TECHNIQUES FOR PREDICTING CHANGES IN THE ECOSYSTEM 153
Kruthi, Anugraha Anil kumar, Aromal A. J. and *Chaya Ravindra*
INTRODUCTION 153
SEISMIC WAVES 154
LITERATURE SURVEY 156
TYPES OF SEISMIC WAVES 158
P-WAVE 158
P waves for Earthquake Cautioning 159
S-WAVE 159
MACHINE LEARNING 159
TYPES OF MACHINE LEARNING 160
Supervised Learning 160
Unsupervised Learning 161
Reinforcement Learning 161
APPLICATIONS OF MACHINE LEARNING 161
DEEP LEARNING 162
APPLICATIONS OF DEEP LEARNING 162
ARTIFICIAL NEURAL NETWORK 163
Feedforward Neural Network 164
Feedback Neural Network 164
CONVENTIONAL NEURAL NETWORK 164
CONCLUSION 165
REFERENCES 165

CHAPTER 11 A DETAILED ANALYSIS OF ISSUES WITH SOLID-STATE DEVICES 168
Shweta and *Hifzan Ahmad*
INTRODUCTION 168
FLASH TRANSLATION LAYER (FTL) 171
GARBAGE COLLECTION 172
Greedy Garbage 173
Cost Benefit (CB) 174
Cost Age Times (CAT) 174
Cost-Age-Time with Age-sort (CATA) 174
Fast and Endurant Garbage Collection (FeGC) 175
Swap-Aware Garbage Collection (SAGC) 176
File-Aware Garbage Collection (FaGC) 176
Efficient and Non-Sensitive File-Aware Garbage Collection (FaGC+) 177
Address Mapping 178
Sector Mapping 178
Block Mapping 180
Hybrid Mapping 180
WEAR LEVELING 181
Evenness-Aware Algorithm 182
Dual Pool Algorithm 182
BAD BLOCK MANAGEMENT 183

Detection of Bad Block ... 183
Replacement of Bad Block ... 183
Skip Block Approach ... 183
Reserve Block Approach (RBA) ... 184
CONCLUDING REMARKS ... 185
REFERENCES ... 185

CHAPTER 12 PROFIT-BASED UNIT COMMITMENT USING LOCAL AND GLOBAL SEARCH METHODS ... 189
Nirmalya Mallick, Ayani Nandi and *Ranadip Roy*
INTRODUCTION ... 189
NOVELTY AND ROAD MAP OF THE CHAPTER ... 190
LITERATURE REVIEW ... 191
PROPOSED METHOD ... 194
PERFORMANCE EVALUATION ... 197
Ten Power Unit System ... 197
Forty Power Unit System ... 200
Hundred Power Unit System ... 202
CONCLUSION ... 208
REFERENCES ... 209

CHAPTER 13 CLASSIFICATION OF DEEP LEARNING TECHNIQUES FOR OBJECT DETECTION ... 212
Aras Amruth Raj Purushotham, Manjunath Ravindra and *Chaya Ravindra*
INTRODUCTION ... 212
Literature Survey ... 216
Two-stage Detectors ... 217
One-stage Detectors ... 217
Deep Learning Models ... 217
CNN ... 218
Convolutional Layer ... 218
Pooling Layer ... 218
Fully Connected Layer ... 219
Dropout Layer ... 219
Activation Functions ... 219
Generic Object Detection ... 219
RCNN ... 219
Selective Search Algorithm ... 221
Problems with RCNN ... 222
Fast RCNN ... 222
Faster RCNN ... 223
CNN Layer ... 224
Anchors ... 224
Object Localization and Classification System ... 224
RoI Pooling ... 224
ONE-STAGE OBJECT DETECTORS ... 225
YOLO ... 225
Residual Blocks ... 225
Bounding Box Regression ... 226
Intersection Over Union (IOU) ... 226
SSD (Single Shot Multibox Detector) ... 226
CONCLUSION ... 226

REFERENCES 227

SUBJECT INDEX 229

PREFACE

An automated machine that performs like a human and replaces the efforts of humans is termed a robot. When these kinds of devices get interlinked with other interdisciplinary technologies for better efficiency, speed, accuracy, output, management and operation, it is termed robotics. Research on the interdisciplinary branches of mechanical devices, data science, the Internet of things, solid-state devices, and others can together make automation more efficient, and it can lead to a revolution in robotics and automation in Industry 4.0. Whenever we say Industry 4.0, it means the fourth Industrial Revolution, which will not only lead the manufacturing sector to a greater height, but will also transform the service and communication sector. Industrial Revolution 4.0 will mostly deal with unmanned vehicles, 3D printing, advanced robotics and new materials. This revolution will support organizational efficiency. Many techniques and areas are designed, such as cloud computing, machine learning, nanobots, supply chain management, information security, *etc*. Hence, the main motive of this book is to explore the application and different technologies associated with robotics and automation in Industry 4.0.

THE OBJECTIVE OF THE BOOK

The objective of this book is to get insights into the tools and technologies of the automation-based Industrial Revolution, which mostly increase organizational efficiency by improving manufacturing, communication and services using new technologies. It also helps to understand different aspects of robotics and automation. It can help different users such as students, research scholars, academicians, industry people, *etc*.

ORGANIZATION OF THE BOOK

The book contains 13 chapters that are organized into four sections as follows. **Section 1** discusses the application and advancement of robotics. **Section 2** highlights the application of renewable energy, which indicates the demand for a sustainable future. **Section 3**discusses the systematic analysis and application of financial technology. **Section 4** discusses some multidisciplinary areas along with techniques and applications.

Section 1: Robotics: Applications and Advancements (Chapters 1-3)

The section discusses different applications and advancements of robotics.

Chapter 1

This chapter discusses different types, principles and applications of nanorobots that are used in different emerging areas. It also discusses the principles of different explored and programmed bots for repairing specific targets.

Chapter 2

This chapter discusses robot path planning systems in a dynamic environment. The Deep Q-learning technique is used in this chapter by using a neural network. It helps to avoid different obstacles in the environment, which are dynamically created by the user.

Chapter 3

This chapter only focuses on innovative design analysis of customized solar frames but not the working of the solar frames or air purifiers. This innovative design is going to change the concept of cycling in the future.

Section 2: Application of Renewable Energy: Power for a Sustainable Future (Chapters 4-5)

The section discusses different renewable energy applications for power management based on a sustainable future.

Chapter 4

This chapter discusses a hybrid optimization technique for profit-based unit commitment. It solves the uncertainty issues for energy source management in wintertime and summertime. It helps reduce several noises that are related to gases that are harmful to fossil fuels and cause diseases and sicknesses.

Chapter 5

This chapter illustrates the impact of Industry 4.0 based on the manufacturing system. The chapter contains several challenges and opportunities based on data production and integrating new technology in the sector of manufacturing.

Section 3: FinTech: Systematic Analysis and its Applications (Chapters 6-8)

The section discusses the systematic analysis of financial technology based on new and innovative applications.

Chapter 6

In this chapter, the authors narrow down the overview of smart financial businesses and their complex challenges. It helps to manage the entire smart FinTech Ecosystem using the fusion of artificial intelligence and data science. It helps to enable smart FinTech and discusses some research problems among global academic and researcher communities.

Chapter 7

This chapter provides insight into blockchain methodology applied in IoT healthcare security. It helps to provide the potential for the medical care environment. It also discusses how customary clinical frameworks and organizations have been occupied with the medical services area throughout the previous years.

Chapter 8

This chapter provides a design development for pharmaceutical applications through the fusion of IoT and blockchain. In this chapter, it is suggested to avoid counterfeit drugs, delivering the pharmaceutical products to customers at the right time, and environmental parameters such as temperature and humidity are also monitored throughout the supply chain to avoid spoilage of pharmaceutical products.

Section 4: Multidisciplinary Approach: Understanding, Benefits and Applications (Chapters 9-13)

The section discusses a different multidisciplinary approach based on new technologies and techniques that help in the automation system.

Chapter 9

This chapter presents a holistic view of attrition and retention of employees on psychological aspects during this cut-throat competitive environment in India. Biology has a little role in management, though one cannot ignore biology in psychology. In broader terms, attrition is somehow related to psychology, and psychology and physiology are two sides of a coin.

Chapter 10

This chapter provides a system for predicting the changes in the ecosystem. The changes in the ecosystem affect the living creatures who depend upon the ecosystem. One of the subsets of machine learning that play a vital role in saving the lives of living creatures is deep learning, which is used in this work for prediction purposes.

Chapter 11

This chapter thoroughly examines solid-state drive subjects, ranging from the physical features of a flash memory cell to the design pattern. The subjects pertaining to the flash translation layer are described within the context of interconnected system-level operations.

Chapter 12

This chapter discusses profit-based unit commitment using the global and local search methods. It suggests the combination of chaotic maps with Harris Hawks Optimizer and chaotic Sine Cosine Algorithm advancement strategy and assesses the execution of the proposed improved technique considering Plug-in Electric Vehicles.

Chapter 13

This chapter discusses the classification of deep learning techniques for object detection. It uses an audit on a machine learning approach for classification. The applications of the protest location have been summarized, along with the diverse approaches to the location of the objects using template-based, portion-based, and region-based methods.

Santosh Kumar Das
Department of Computer Science and Engineering at Sarala Birla University, India

Soumi Majumder
Department of Business Administration, Future Institute of Engineering and Management, India

&

Nilanjan Dey
Department of Computer Science and Engineering at Techno International New Town, India

List of Contributors

Anugraha Anil kumar	School of ECE, REVA University, Bengaluru, India
Aromal A. J.	School of ECE, REVA University, Bengaluru, India
Ayani Nandi	Department of Electrical Engineering, Sanaka Educational Trust's Group of Institutions, Durgapur, India
Arun Kumar	National Institute of Technology, Patna, Bihar-800005, India
Abhisek Sarkar	Dinabandhu Andrews Institute of Technology and Management, kolkata, India
Chaya Ravindra	School of ECE, REVA University, Bengaluru, India
Hiren Joshi	Gujarat University Ahmedabad, Gujarat, India
Hifzan Ahmad	Dr. A.P.J Abdul Kalam Technical University, Lucknow, India
Jyothy S. T	REVA University, Bangalore, India
Kruthi	School of ECE, REVA University, Bengaluru, India
Keyurkumar Patel	School of IT, Artificial Intelligence and Cyber Security, Rashtriya Raksha University, Gandhinagar, Gujarat-382305, India
Manorama Patnaik	Amity Institute of Information Technology, Amity University, harkhand, Ranchi, India
Mrinal Sarvagya	REVA University, Bangalore, India
Nirmalya Mallick	Department of Electrical Engineering, Sanaka Educational Trust's Group of Institutions, Durgapur, India
Preeti Chandrakar	Department of CSE, National Institute of Technology Raipur, India
Pujita Sunnapu	Unitedworld School of Computational Intelligence, Karnavati University, Uvarsad, Gandhinagar–382422, India
Parkarsh Kumar	Sarala Birla University, Ranchi, India
Ramesh Chandra Goswami	Gujarat University Ahmedabad, Gujarat, India
Ranadip Roy	Department of Electrical Engineering, Sanaka Educational Trust's Group of Institutions, Durgapur, India
Rifaqat Ali	Department of Mathematics and Scientific Computing, National Institute of Technology Hamirpur, H.P., India
Santosh Kumar Das	Sarala Birla University, Ranchi, India
Suraj	National Institute of Technology, Patna, Bihar-800005, India
Sunil Gautam	Department of Computer Science and Engineering, Institute of Technology, Nirma University, Ahmedabad, Gujarat-382481, India
Shweta	Madan Mohan Malaviya University of Technology, Gorakhpur, Uttar Pradesh, India
Tameshwer Nath	Department of mechatronics, IIIT Bhagalpur, Bihar-813210, India

CHAPTER 1

Nanorobots: Types, Principles and Applications

Parkarsh Kumar[1] and **Santosh Kumar Das**[1,*]

[1] *Sarala Birla University, Ranchi, India*

Abstract: Nanobots or Nanorobots are one of the emerging applications in both nanotechnology and robotics. These bots are programmed to carry out specific applications for a specific purpose. Owing to their properties such as smaller volume, efficiency and accuracy, nanobots are being explored in different fields of study, especially nanomedicine, automation, drug delivery, chemistry, aerospace and others. These bots can be programmed and explored in such a way that they can be used to repair the specific target in the body, which is impossible using bare hands. In this chapter, we are going to explore such types of applications and their principles.

Keywords: Application of nanobots, Nanobots, Types of nanobots.

INTRODUCTION

In 1959, Professor Richard Feynman, through his talk "Plenty of Room at the Bottom", laid the fundamentals of nanotechnology [1]. In one of the chapters, he discussed molecular machines, the application of which includes nanorobotic surgery, drug delivery, and others. As suggested by Professor Feynman, the miniaturization of robotics has been explored, and advancements have been made in this field over time. These are types of nanodevices whose application can be explored in many fields, but currently, it is mostly being explored in the field of biomedicals. One of the first studies on nanobots was conducted by Robert Freitas, which is based on the principle of molecular manufacturing [2]. This nanodevice was created to function like an artificial erythrocyte.

Compared to natural red cells, these artificial devices were capable of transporting a similar amount of carbon dioxide and 236 times more oxygen to tissues. These devices were fabricated using hard diamondoid materials and consist of different types of chemicals, pressure and thermal sensors to monitor complex responses and behavior. Since then, nanobots have evolved in multidimensional fields with increasing applications. Mavroidis *et al.* [3] explained the evolution of fully

* **Corresponding author Santosh Kumar Das:** Sarala Birla University, Ranchi, India;
E-mail: sunsantosh2014@gmail.com

functional autonomous nanorobots and the integration of different components into a system. Owing to its small size, durability, self-replication and mobility, multiple applications of nanobots have been developed to solve problems related to the environment, agriculture, biology and medicine. The design of different nanorobots is inspired by different kinds of species present in the environment to perform specific operations. These nanobots seem to be very promising in areas of invasive surgery, drug delivery, contamination screening, controlling pests and plant viruses, antimicrobial activity, breaking blood clots, parasite removals, and others [4 - 6]. In order to scale up, researchers across different organizations are trying to overcome limitations in terms of high-quality design, cost analysis, biocompatibility, interfacing with blood, which is viscous, and control issues. Overcoming the challenges will help in achieving unlimited applications especially in the field of biomedical. The current trend of nanorobots in precision medicine, which includes surgeries, therapy, imaging and diagnosis, is being studied [7]. These bots will have the ability to locate specific positions in the body and deliver the precise dosage. With noninvasive medication and various groundbreaking innovations, it is expected that the market of nanobots will exponentially increase to 21.45 billion dollars in 2030 from its present value of 6.19 billion dollars.

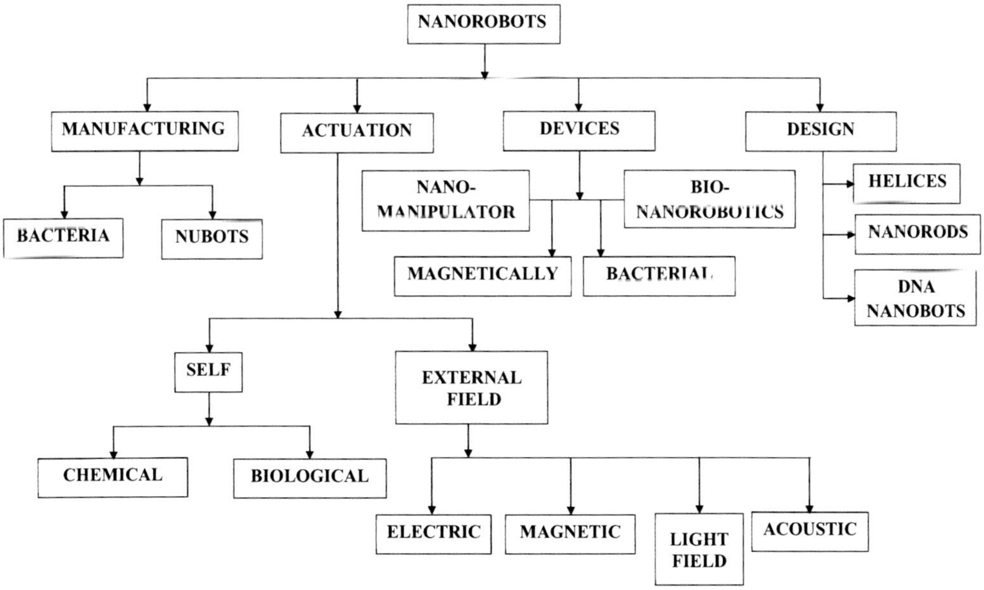

Fig. (1). Types of nanorobots.

TYPES

Further, we are going to discuss different types of nanobots based on a few parameters such as design, nanorobotic devices, actuation types and manufacturing approaches. These are just a few parameters, which can be further subdivided into many categories. Some types of robots based on the subdivision of parameters can be similar in nature. Nanobots classification and their respective subcategories are shown in Fig. (**1**).

- On the basis of design: Based on the applications, many designs of nanobots have been proposed, but they are generally classified into three categories on the basis of design. Arvidsson *et al.* described them as Helices, DNA Nanobots and Nanorods [8 - 11]. They are discussed in detail in further sections.
- On the basis of actuation: In general, the movement of nanobots takes place in a medium having a low Reynolds number, where inertial force is negligible as compared to viscous force. This makes it difficult to operate, and hence nanobots need to be uninterruptedly powered for actuation. Since it is very difficult to load energy sources on nanorobots, researchers have started focusing on actuation technologies for nanorobotics. Xu *et al.* [12] summarized different types of actuation techniques into two sub-categories, which are self-actuation and external field actuation. In order to actuate nanobots using a variant magnetic field, as explained by Dreyfus *et al.* [13], it is subject to get a constant and oscillating magnetic field along the horizontal and perpendicular direction, respectively. Later, many other techniques, such as U-shaped, fish-like structure, freestyle swimmer, sea star and other types of magnetically actuated nanobots, have been proposed for different applications. Although there is a safety issue related to high-intensity magnetic fields, strong pungent power and no harm to the biological system make it viable to experiment on. We know that an electric and magnetic field can be transformed into each other, so robotic systems with certain conditions and without complicated mechanical parts were proposed, which could utilize an external electric field for actuation. For actuation, some researchers use the finite element technique to produce a magnetic field using coil assembly. The concept of quad nanopore devices is also used for the purpose of actuation. Other types of external field actuation include light field actuation and acoustic actuation. Experiments were conducted to establish a relationship between laser intensity and the speed of nanoswimmers for biomedical and soft matter applications. Based on such principles, needle-shaped nanoswimmers, electrochemical actuators, and photoresponsive DNA signaling molecules have been designed over time. On the other hand, researchers used the principle of ultrasonic bubble-gathering effect for the movement of robots. Rod-shaped nanobots that can rotate, arrange, suspend and

assemble in the ultrasonic range are fabricated. Some researchers achieve cell manipulation and actuation using acoustic radiation force in standing waves. In the case of self-actuation methods, the static balance of nanorobots gets disturbed for the purpose of movement. Before breaking the static balance, an asymmetrical chemical field needs to be created. Whitesides *et al.* [13] was one of the first researchers who performed self-actuation in robots by converting chemical energy into kinetic energy using the principle of oxidation-reduction. Later, tubular and platinum-based V-shaped robots were developed for self-actuation. Some researchers even worked on a hybrid Au/Pt (z-shaped) self-actuation system for targeted drug delivery. These chemical self-actuation devices were biocompatible, but these methods also have safety issues, short action time and nontoxic urea. Biological self-actuation can be used to overcome the issue of safety. These types of nanobots have been developed by taking inspiration from nature. Flagella and Magnetotactic bacteria give promising results for actuating force and smart nanobots. Droplets of different bacteria were placed in a solution of glucose and water to generate movement with the help of ethylenediaminetetraacetic acid [14].

- On the basis of devices: Nanobots can be broadly classified into four categories. Mavroidis *et al.* [3] mentioned these four systems as Nanomanipulators, BioNano Robotics, Magnetically Guided Nanorobotics and Bacteria-based Nanorobotics. The microscopic viewing system that helps in viewing small-scale objects is termed the nanomanipulator. Scanning probe microscopes are used to enlarge images, which helps in a better understanding of devices and manipulation at atomic scales. Some biological elements, such as DNA and proteins, are capable of generating force or signal at the cellular level. Devices made from these elements are called nanocomponents, and the robotic system is called the bio nanorobotic system. Developed in 2003, these systems are now capable of transformation into motors, sensors, joints and other components of robotic systems. It has been hard to develop full-scale nanorobots, but with the development of the ferromagnetic material-based magnetically guided nanorobotic system, nanobots with six degrees of freedom (DOF) can be realized and external magnetic force can be used for propulsion and actuation. Another type, called bacteria-based nanorobotic system, is developed by taking information from nature [15]. Although it has similarities with both magnetic and bio nanorobotic systems, it differs in terms of design, control and guidance.
- On the basis of manufacturing: Based on the approaches for manufacturing nanorobots, they can either be classified as NUbots or bacteria-based [16]. The bacteria-based approach uses flagellum and electromagnetic field for propulsion and control of motion by taking inspiration from microorganism bacterium such as *E. coli*. It is the same as what we have discussed in other classifications of nanobots. Nubots are also called nucleic acid robots. They are a type of

molecular machine thatis fabricated and activated using molecules, proteins and DNA.

These are the limited classifications. Further subcategories of nanobots can be created by conducting a comparative study on nanobots on the basis of different parameters related to them.

Applications and Recent Developments

We have discussed different types of nanobots, their principle and their fabrication approaches. It is yet to be fully scaled, but based on the approaches, it has been slightly used in different parts of devices, especially in biomedical devices. Experts from different fields of study came together to make advancements in nanobots, which is certainly the future of precision medicine, non-invasive diagnostics, environmental science, agriculture, military, space and others. Most of the devices are still in the preliminary phase of development, and lots of research and trials need to be performed in order to make them commercially and economically viable.

With research advancement and commercialization, full-scale, economically viable nanobots can be developed to solve problems related to medicine, like cancer treatment, by replacing damaged neurons or by transferring specific drugs to specific sites, thereby safeguarding patients from the harmful process of chemotherapy [4] [13]. It can also be used in the body to monitor abnormalities and provide noninvasive ways to treat them. Using this kind of surveillance technique, it can help in checking the contents of blood and provide warning about any kind of possible diseases. Other medical problems that can be monitored and diagnosed are blood clots, body clots, diabetes, gout, and others.

With rapid development in the field of materials and biochemistry, the future of nanobots is near, but challenges remain in terms of removal of toxicity, proper miniaturization of devices, compatibility, energy harvesting, positional nano assembly, noise interference and dynamics of robotic systems [17]. Some decision making, AI-powered tools and machine learning can also be integrated in the future for better outcomes [18 - 23]. Some challenges are also present during the designing of nanorobots, such as high complexity, interface design, cost, *etc.* There are several post-quantum cryptography techniques available to implement security in the nanorobots [24 - 27]. They help to reduce several attacks in the context of different applications. Lightweight cryptography also helps to protect against some attacks [28 - 34].

CONCLUSION

The need to describe rarefied flows in a variety of applications that involve micro and nanoscale pores or channels has stimulated the development of a multitude of theoretical and numerical methodologies. A comparison of the predictions of the mesoscopic approaches with those of the Navier-Stokes formulation complemented by a non-slip flow condition revealed that the latter can be safely employed for very small Knudsen number values only. The DSMC method was used to determine the flow field in different types of porous media and processes.

REFERENCES

[1] R.P. Feynman, "Plenty of room at the bottom", *Eng. Sci.,* vol. 23, no. 5, pp. 22-36, 1959.

[2] R.A. Frcitas, "Exploratory design in medical nanotechnology: A mechanical artificial red cell", *Artif. Cells Blood Substit. Immobil. Biotechnol.,* vol. 26, no. 4, pp. 411-430, 1998. [http://dx.doi.org/10.3109/10731199809117682] [PMID: 9663339]

[3] C. Mavroidis, and A. Ferreira, "Nanorobotics: Past present and future", In: *Nanorobotics.* Springer-Verlag: Berlin, Germany, 2013, pp. 3-27. [http://dx.doi.org/10.1007/978-1-4614-2119-1_1]

[4] M.T. Nistor, and A.G. Rusu, "Nanorobots with applications in medicine", In: *Polymeric Nanomaterials in Nanotherapeutics* Elsevier, 2019, pp. 123-149. [http://dx.doi.org/10.1016/B978-0-12-813932-5.00003-0]

[5] Y. Saadeh, and D. Vyas, "Nanorobotic applications in medicine: Current proposals and designs", *Am. J. Robot. Surg.,* vol. 1, no. 1, pp. 4-11, 2014. [http://dx.doi.org/10.1166/ajrs.2014.1010] [PMID: 26361635]

[6] A. Cavalcanti, W.W. Wood, L.C. Kretly, and B. Shirinzadeh, "Computational nanorobotics: Agricultural and environmental perspectives", *Nanomedicine,* vol. 2, pp. 82-87, 2003.

[7] F. Soto, J. Wang, R. Ahmed, and U. Demirci, "Medical micro/nanorobots in precision medicine", *Adv. Sci.,* vol. 7, no. 21, p. 2002203, 2020. [http://dx.doi.org/10.1002/advs.202002203] [PMID: 33173743]

[8] F. Qiu, and B.J. Nelson, "Magnetic helical micro and nanorobots: Toward their biomedical applications", *Engineering,* vol. 1, pp. 021-026, 2015.

[9] H. Zhou, C.C. Mayorga-Martinez, S. Pané, L. Zhang, and M. Pumera, "Magnetically driven micro and nanorobots", *Chem. Rev.,* vol. 121, no. 8, pp. 4999-5041, 2021. [http://dx.doi.org/10.1021/acs.chemrev.0c01234] [PMID: 33787235]

[10] Y. Xu, Q. Bian, R. Wang, and J. Gao, "Micro/nanorobots for precise drug delivery *via* targeted transport and triggered release: A review", *Int. J. Pharm.,* vol. 616, p. 121551, 2022. [http://dx.doi.org/10.1016/j.ijpharm.2022.121551] [PMID: 35131352]

[11] Z. Wang, Z. Xu, B. Zhu, Y. Zhang, J. Lin, Y. Wu, and D. Wu, "Design, fabrication and application of magnetically actuated micro/nanorobots: A review", *Nanotechnology,* vol. 33, no. 15, p. 152001, 2022. [http://dx.doi.org/10.1088/1361-6528/ac43e6] [PMID: 34915458]

[12] K. Xu, and B. Liu, "Recent progress in actuation technologies of micro/nanorobots", *Beilstein J. Nanotechnol.,* vol. 12, pp. 756-765, 2021. [http://dx.doi.org/10.3762/bjnano.12.59] [PMID. 34367859]

[13] R. Dreyfus, J. Baudry, M.L. Roper, M. Fermigier, H.A. Stone, and J. Bibette, "Microscopic artificial swimmers", *Nature,* vol. 437, no. 7060, pp. 862-865, 2005. [http://dx.doi.org/10.1038/nature04090] [PMID: 16208366]

[14] L. Wang, A.C. Hortelão, X. Huang, and S. Sánchez, "Lipase-powered mesoporous silica nanomotors for triglyceride degradation", *Angew. Chem. Int. Ed.,* vol. 58, no. 24, pp. 7992-7996, 2019. [http://dx.doi.org/10.1002/anie.201900697] [PMID: 30990243]

[15] B. Behkam, and M. Sitti, "Bacterial flagella-based propulsion and on/off motion control of microscale objects", *Appl. Phys. Lett.,* vol. 90, no. 2, p. 023902, 2007. [http://dx.doi.org/10.1063/1.2431454]

[16] SS Upadhye, BK Kothali, AK Apte, AA Kulkarni, VS Khot, AA Patil, and RN. Mujawar, "A review ON nanorobots", *Am. J. PharmTech Res.,* vol. 9, no. 2, pp. 1-20, 2019.

[17] R. Padshala, V. Rajan, and P. Patani, "Nanobots: Future and Development", *J. Pharm. Negat. Results,* vol. 7, pp. 1967-1975, 2022. [http://dx.doi.org/10.47750/pnr.2022.13.S03.293]

[18] C. Decision-Making, *Systems in Engineering.* IGI Global, 2023.

[19] *Constraint Decision-Making Systems in Engineering* IGI Global, 2022.

[20] *AI-empowered knowledge management* Springer, 2021.

[21] "A Non-Linear Multi-Objective Technique for Hybrid Peer-to-Peer Communication", *Inf. Sci.,* 2023.

[22] "An optimal data communication system in wireless sensor network using artificial neural network", In: *Constraint Decision-Making Systems in Engineering* IGI Global, 2023, pp. 120-134.

[23] *Data-Driven Approach for Bio-medical and Healthcare* Springer Nature, 2022.

[24] A. Sarker, "Error detection architectures for hardware/software co-design approaches of number-theoretic transform", *IEEE Trans. Comput. Aided Des. Integrated Circ. Syst.,* 2022.

[25] M.M. Kermani, and R. Azarderakhsh, "Integrating emerging cryptographic engineering research and security education", *2015 ASEE Annual Conference & Exposition,* pp. 26-989, 2015.

[26] M. Mozaffari-Kermani, "Reliable and high-performance hardware architectures for the Advanced Encryption Standard/Galois Counter Mode", In: *Doctoral dissertation, The University of Western Ontario, Canada,* 2011.

[27] A. Sarker, M.M. Kermani, and R. Azarderakhsh, "Efficient error detection architectures for postquantum signature falcon's sampler and KEM SABER", *IEEE Trans. Very Large Scale Integr. (VLSI) Syst.,* vol. 30, no. 6, pp. 794-802, 2022.

[28] A.C. Canto, J. Kaur, M.M. Kermani, and R. Azarderakhsh, "Algorithmic security is insufficient: A comprehensive survey on implementation attacks haunting post-quantum security", *arXiv:2305.13544,* 2023.

[29] R. Azarderakhsh, B. Koziel, S. F. Langroudi, and M. M. Kermani, "FPGA-SIDH: High-performance implementation of supersingular isogeny Diffie-Hellman key-exchange protocol on FPGA", *Proc. eprint,* vol. 672, pp. 1-18, 2016.

[30] E. Dubrova, K. Ngo, J. Gärtner, and R. Wang, "Breaking a fifth-order masked implementation of crystals-kyber by copy-paste", *Proceedings of the 10th ACM Asia Public-Key Cryptography Workshop,* pp. 10-20, 2023. [http://dx.doi.org/10.1145/3591866.3593072]

[31] J. Kaur, A.C. Canto, M.M. Kermani, and R. Azarderakhsh, "A comprehensive survey on the implementations, attacks, and countermeasures of the current NIST lightweight cryptography standard", *arXiv:2304.06222,* 2023.

[32] M. M. Kermani, R. Azarderakhsh, and J. Xie, "Error detection reliable architectures of Camellia block cipher applicable to different variants of its substitution boxes", *IEEE Asian Hardware-Oriented Security and Trust (AsianHOST),* pp. 1-6, 2016. [http://dx.doi.org/10.1109/AsianHOST.2016.7835560]

[33] A. Cintas-Canto, J. Kaur, M.M. Kermani, and R. Azarderakhsh, "ChatGPT vs. lightweight security: First work implementing the NIST cryptographic standard ASCON", *arXiv:2306.08178,* 2023.

[34] M.M. Kermani, E. Savas, and S.J. Upadhyaya, "Guest editorial: Introduction to the special issue on emerging security trends for deeply-embedded computing systems", *IEEE Trans. Emerg. Top. Comput.,* vol. 4, no. 3, pp. 318-320, 2016. [http://dx.doi.org/10.1109/TETC.2015.2482778]

CHAPTER 2

Robot Path Planning in a Dynamic Environment Using Deep Q-Learning

Rifaqat Ali[1,*] and **Preeti Chandrakar**[2]

[1] *Department of Mathematics and Scientific Computing, National Institute of Technology Hamirpur, H.P., India*

[2] *Department of CSE, National Institute of Technology Raipur, India*

Abstract: Robot path planning is a necessary requirement for today's autonomous industry as robots are becoming a crucial part of the industry. Planning a path in a dynamic environment that changes over time is a difficult challenge for mobile robots. The robot needs to continuously avoid all the obstacles in its path and plan a suitable trajectory from the given source point to the target point within a dynamically changing environment. In this study, we will use Deep Q-Learning (Q-Learning using neural networks) to avoid the obstacles in the environment, which are being dynamically created by the user. The main aim of the robot is to plan a path without any collision with any of the obstacles. The environment is simulated in the form of a grid that initially contains information on the starting and the target location of the robot. Robots need to plan an obstacle-free path for the given points. The user introduces obstacles whenever he/she wishes during the simulation to make the environment dynamic. The accuracy of the path is judged by the path planned by the robot. Various architectures of neural networks are compared in the study that follows. Simulation results are analyzed for the evaluation of an optimized path, and the robot is able to plan a path in the dynamic environment.

Keywords: Artificial neural network, Deep learning, Neural networks, Reinforcement learning, Robot.

INTRODUCTION

Nowadays, with the advancement of technology and a tremendous boom in the field of robotics, robots are taking over most mundane tasks that do not require much cognitive ability and can be performed easily by unskilled laborers. In 2010, Chen *et al.* [1] showed that robots used to serve and cater to customers in a human-interactive environment in restaurants. Later, Chen *et al.* [2] also proposed

* **Corresponding author Rifaqat Ali:** Department of Mathematics and Scientific Computing, National Institute of Technology Hamirpur , H.P., India; E-mail: rifaqatali27@gmail.com

Santosh Kumar Das, Soumi Majumder and Nilanjan Dey (Eds.)

the vacuum robots used for cleaning. In 2015, multiple robots communicated with each other over a shared network and carried loads from one fixed location to another fixed location, as shown by Das *et al.* [3]. Autonomous mobile robots have found their way into industries, military research programs, and many scientific research programs. The Schemes [4 - 6] present additional information on robots used in human-friendly environments like education, entertainment, surveillance and many more. As seen, robots continuously keep evading new human-centric and non-human-centric environments to find vast applications. Robots need to be as autonomous as possible *i.e.*, they should possess the ability to navigate, find paths and perform various activities without the intervention of any kind to accomplish all such tasks. All the activities that are performed by autonomous mobile robots require the design and implementation of a path planning and path optimization algorithm at its core.

Robot path planning, being at the crux, plays a vital role in the design, implementation and manufacturing of autonomous robots. If implemented effectively, path planning significantly improves the following parameters in an autonomous robot:

Robot Accuracy: The algorithm must produce next to zero errors because the robot is expected to work in human-friendly environments where the slightest errors can be catastrophic. Hence, there is a need to test this out and make sure the robot is accurate enough that it stands firm on its capabilities.

Task Repeatability: Once the implementation of the path planning algorithm is complete, including the training phase of the robot, it can perform the same task hundreds of times with high accuracy and effectiveness without any slack.

Product Quality: Based on accuracy and repeatability, we can directly conclude that our robot has been trained well and can assume a result that will depict the high-quality work of our robot.

One can define path planning as the continuous locomotion of a robot from the initial/starting configuration to the final/goal configuration in an unknown environment. The environment in which the robot moves consists of free space and obstacles. The robot needs to ensure that it does not collide with any obstacle during its course. Path optimization deals with finding the path with the least cost and distance and which consumes the least amount of time. Fig. (**1a**) shows an example environment with obstacles and defined configurations for start and goal states. Fig. (**1b**) shows an example of a path planned in the shown environment.

Structure of the Paper

This article is organized into 7 sections. Following the introduction, Section 2 presents different approaches and work, which were used by various researchers in the past to solve the problem. Also, the section ends with a description of the approach used in this article. Section 3 of the article describes the methodology used to solve the problem in addition to the nuances of the algorithm. Section 4 discusses the experimental setup and the various technologies used, along with the various parameters of the algorithm used during the experiment. Section 5 analyses the results and the performance of the experiment with the help of some graphs. The article comes to an end by presenting a conclusion for the experiment in Section 6.

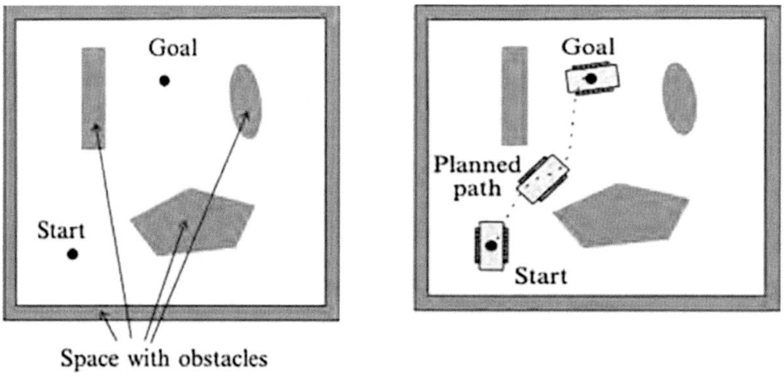

Fig. (1). (a) Environment with obstacles and start and goal configurations, and (b) one out of many possible paths from the start to the goal configuration.

RELATED WORKS

In recent decades, many approaches have been devised by researchers to find a solution to the problem of path planning and optimization for autonomous mobile robots. These approaches range from the use of graph-based algorithms to the use of contemporary neural networks to implement deep learning and reinforcement learning algorithms. Some of these approaches and algorithms have been briefly described as follows:

Graph-Based Approach

In this type of approach, the environment is considered as a connected network consisting of nodes and edges similar to a graph data structure. The nodes that constitute the obstacles are considered forbidden since the robot is not allowed to use these nodes while ascertaining a path from the initial to the goal state. All other nodes form the free space of the environment and are traversable as the robot moves. Classical graph traversing algorithms, such as the Breadth-First Search (BFS) and the Depth First Search (DFS), are fruitful in determining solutions to the path planning problem. Single-source shortest path algorithms, such as the Dijkstra algorithm and the Bellman-Ford algorithm, are also capable of determining the path from the source node (initial state) to the destination node (goal node). Terzimehic *et al.* [7] successfully used the algorithms mentioned above to find solutions to the problem.

The major drawback of the algorithms stated above was (1) the amount of time they consumed as the size of the environment increased and (2) all the nodes in the environment had to be traversed before the most optimal path could be determined. Heuristic approaches overcame such drawbacks to an extent and enhanced the performance of traditional algorithms. A heuristic at a node can be defined as a function or attribute that the algorithm uses to determine which neighboring node is most likely to result in an optimized path to the goal node. The most famous heuristic-based algorithm is the A* [8]. Many algorithms like the Iterative Deepening A* (IDA*) [9], Anytime Repairing A* (ARA*) [10], Jump Point Search (JPS) [11], Theta* [12] *etc.*, which are modified versions of the A* algorithm, fall under the category of heuristic search algorithms. Nature-inspired heuristic algorithms such as the Genetic Algorithm (GA) [13, 14], Artificial Bee Colony (ABC) [15], Cuckoo Search Algorithm [16], and Firefly Algorithm (FA) [17] have been researched and developed over decades and fall under the category of heuristic and meta-heuristic algorithms.

Fuzzy Logic Approach

In 1965, Prof. Zadeh [18] laid the foundation for fuzzy sets. Fuzzy logic governs the implementation of fuzzy sets, which provide a mathematical notation for the representation of fuzzy logic. Traditional logic (also known as binary logic) tries to evaluate problems based on two discrete truth values, namely 'true' and 'false' (0 or 1). The major drawback of traditional logic is that it cannot account for uncertainty in real-world problems such as linguistics. Fuzzy logic overcomes this drawback by defining a membership function that determines the degree of membership of each element in the fuzzy set. The membership values of the elements in the set can be anywhere in the range of 0 and 1. The process of converting a crisp set to a fuzzy set is called fuzzification, and the process of converting a fuzzy set back to a crisp set is called defuzzification. The problem of path planning using fuzzy logic is solved mostly by dividing the problem into simpler and smaller sets of tasks, where actions planned to achieve a defined set of objectives are based on fuzzy rule statements.

Later, in 1986, Vachtsevanos and Hexmoor [19] suggested the idea of using fuzzy logic to solve the problem of autonomous mobile robot path planning. A set of rules described with the help of the if-then structure can be phrased using fuzzy logic, which in turn controls the decision-making process of the robot path planning algorithm in a dynamic environment. Heading into the year 2013, Chang and Taeseok [20] used a fuzzy inference model along with the process of fuzzification, defuzzification, and sensor fusion to search for the goal state in an unknown dynamic environment. Valdez *et al.* [21], in 2014, used fuzzy logic to adapt specific parameters to optimize the path planning algorithms. Optimization methods mainly cover algorithms such as Particle Swarm Optimization, Ant Colony Optimization, *etc.*, which utilize fuzzy logic to enhance performance and efficiency. Many other fuzzy logic-based solutions are described in other studies [22 - 24].

Artificial Neural Network-Based Approach

Artificial Neural Networks (referred to as Neural Networks later in the text) have been inspired by the biological neural networks present in the human brain. Neural Networks are complex networks of neurons that are capable of processing data at high speeds and identifying relationships between data that are not easily perceivable. These relationships present among the data help establish patterns, which can be used to generalize many solutions to real world problems. Of late, there has been an increasing application of neural networks in forecasting, financial services, fraud detection, stock market price prediction, *etc.*

In 1988, Zacksenhouse *et al.* [25] used Neural Networks to perform cue-based motion planning. This laid the foundation for Neural Networks being used for motion planning applications. In 2010, Nichols *et al.* [26] used a Spiking Neural Network for robot navigation along a wall. Antonelo *et al.* [27] showed the use of a learning process using Recurrent Neural Networks (RNN) in the same year to devise goal-oriented robot navigation. In 2011, Chi *et al.* [28] managed to make a robot traverse through a maze by using Neural Networks to recognize maze patterns and avoid arbitrary obstacles. Motlagh *et al.* [29], in 2014, presented a technique based on reinforcement learning using Neural Networks to enable a mobile robot to learn and navigate an environment constructed by experts. As seen, Neural Networks have been used time and again to tackle the problem of path planning and optimization and continue to be a leading front for research.

Many other approaches apart from the ones discussed above have been presented and used to effectively solve the path planning, optimization and obstacle avoidance problems in the case of mobile robots [30]. A study described some significant approaches like Voronoi Diagrams, Artificial Immune Systems (AIS), Statistical Methods and Wireless Sensor Networks.

In this article, an Artificial Neural Network-based approach using a Deep Q-Learning Algorithm to simulate the path planning and optimization of an autonomous mobile robot is presented. The Deep Q-Learning Algorithm is an advancement over the Q-Learning Algorithm, which is a type of reinforcement learning algorithm. Using a reinforcement learning approach, the pre-requisite of having a defined dataset for training the Neural Network is avoidable. Hence, one does not require a pre-constructed map of the environment in which the obstacles have been defined beforehand. Therefore, during the simulation presented in this article, the context of the environment, such as the inclusion of additional obstacles, changing the shape of obstacles, changing the goal configuration, *etc.*, is easily achievable. This gives a lot of flexibility to the end user of the autonomous robot in terms of re-configuring the robot each time there is a change in the environment. Also, the learning time of the robot can be drastically reduced by using an efficient neural network architecture, which can easily be configured according to the working environment and the demand of the end user.

PROPOSED METHODOLOGY

As stated above, this article utilizes the concept of Deep Q-Learning, which is derived from the application of Deep Learning using Neural Networks to implement a reinforcement learning algorithm known as Q-Learning. The following subsections try to briefly describe the major terminologies and concepts associated with reinforcement learning, Q-Learning and Deep Q-Learning.

The Agent

An entity responsible for performing desired operations using a defined set of actions in an environment is called an agent. The agent learns to perform the desired functions in the most optimized way by exploring the environment. Every iteration in which the agent completes the desired operation is called an episode. The training phase of the reinforcement learning algorithms requires many such episodes.

In the context of this article, the autonomous robot is the agent. The desired function of the autonomous robot is to independently explore the environment and learn to reach a goal configuration from the start configuration. The path planned by the robot should avoid all the obstacles in the environment and should be the most optimized in terms of time taken and distance covered.

The Environment

The region consisting of the free space and obstacles in which the autonomous robot functions is called the environment. The environment may be finite, if the boundaries of the environment are defined, or it may be infinite. The start configuration, goal configuration and obstacles are defined in the environment. The environment can be perceived by a robot as a coordinate system, as a set of nodes forming a graph data structure, as Voronoi segments forming a Voronoi map, *etc.*

In this article, we consider a finite environment represented by a two-dimensional array corresponding to the pixels on the GUI window of the simulator. The goal and start state are pre-defined. The obstacles in the environment are drawn by the user on the simulator window using a mouse or a graphics tablet similar to a paint application in the Windows operating system. The obstacles can be of any shape and size; additional obstacles can be drawn during any phase of the simulation. The elements of the matrix corresponding to the obstacle pixels are not allowed to be traversed by the autonomous robot.

Reinforcement Learning

Reinforcement learning is a technique in which the agent explores the environment and gains all kinds of experiences; the agent learns from these experiences to perform the desired function optimally. Reinforcement learning uses the concept of rewards. Each time the agent changes its configuration by performing an action, it gains a reward. The rewards can be positive or negative based on good and bad experiences, respectively. The overall goal of the agent is to maximize the reward over time. Hence, each episode over time is better than the previous in terms of optimality. Fig. (**2**) shows a block diagram which describes this scheme.

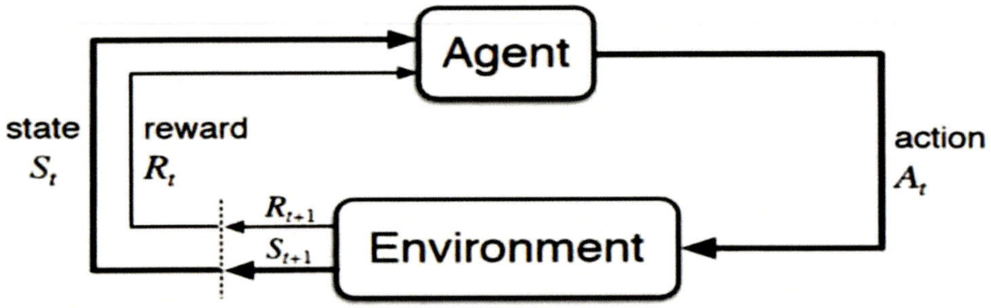

Fig. (2). Block Diagram showing Reinforcement Learning Scheme.

The scheme can be mathematically represented by using the Bellman Equation. In the most simplistic form, the Bellman Equation for a deterministic environment can be expressed as follows:

$$V(s) = \max_{a} (R(s,a) + \gamma V(s')) \quad \textbf{(1)}$$

And for a stochastic environment, the equations can be expressed as:

$$V(s) = \max_{a} (R(s,a) + \gamma \textstyle\sum_{s'} P(s,a,s')V(s')) \quad \textbf{(2)}$$

Where, in Eqs. (1) and (2),

s ≡ current state of the agent

s′ ≡ next state of the agent

a ≡ current action

γ ≡ discount factor ($0 \leq \gamma \leq 1$)

V (s) ≡ value of reward at current state

V (s′) ≡ value of reward at the next state

R(s, a) = reward gained by transitioning from s to s′ using a

P (s, a, s′) ≡ probability of transitioning from s to s′ using a

These equations are a crude version of the study presented in 'The Theory of Dynamic Programming' by Richard Bellman [31]. As backed by the Bellman equations, Reinforcement Learning works on past history of experiences. As the experiences increase, storing all the past experiences and making a decision based on them would require a large amount of memory and processing power. The solution to this conundrum lies in the Markov Decision Process, which is described below.

Markov Decision Process

Markov Decision Process can be stated as a memoryless random process. A stochastic process has the Markov property if the conditional probability distribution of the future states of the process (conditional on both past and present states) depends only upon the present state, not on the sequence of events that preceded it. A process with this property is called a Markov process. Considering the states of the environment as S and the transition from one state to another as P, the tuple can be written as (S, P), as shown in Fig. (**3**).

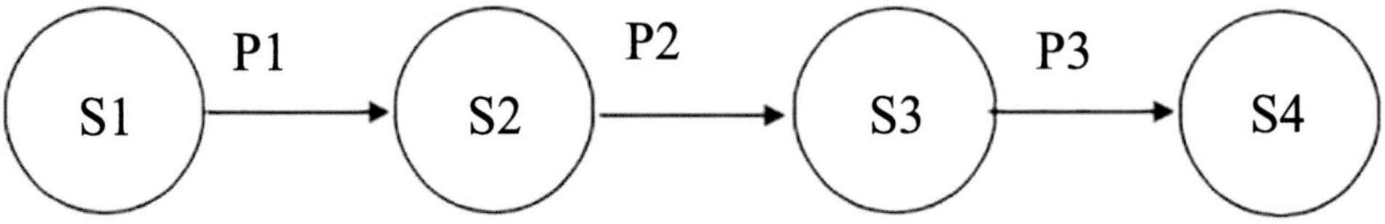

Fig. (3). Markov Process Flow Diagram.

This process provides a mathematical framework for modelling decision making in situations where the outcomes are partly random and under the control of the decision maker. Mathematically, the Markov Decision Process (MDP) can be defined in the form of a tuple as follows:

$$S, A, P_{sa}, \gamma, R\} \quad \textbf{(3)}$$

Where, in Eq. (3),

S ≡ set of states

A ≡ set of actions

P_{sa} ≡ probability of the state transition

γ ≡ discount factor ($0 \leq \gamma \leq 1$)

R ≡ reward function

Q-Learning

Q-Learning is a basic form of Reinforcement Learning where learning is based on the Q-values instead of the values assigned to each of the individual states, as shown in Eq. (2). Q-values show the quality of each action from the current state. For every action a in the set of actions A, and for every state s in the set of states S, we update the Q-value Q, which depends on the reward R received by the agent by taking an action a from the current state s to move to the next state s′. Hence, mathematically, Q-values are defined by the equation shown below:

$$Q(s,a) = R(s,a) + \gamma \Sigma_{s'}(P(s,a,s')V(s')) \quad \textbf{(4)}$$

Now, using Eq. (2), we can modify Eq. (4) as

$$Q(s,a) = R(s,a) + \gamma \sum_{s'} (P(s,a,s') \max_{a} Q(s',a')) \quad \textbf{(5)}$$

where Q(s, a) denotes the Q-value obtained from current state s by taking action a; Q(s′, a′) denotes the Q-value for the next state s′ and next action taken a′ and all the rest of the symbols have the usual meaning defined above.

A table for Q-values corresponding to each state and action pair is maintained and frequently updated based on experiences and rewards. This table is used to decide the next action to be taken to reach an optimal solution. The updation of Q-values is based on the following expression:

$$Q_t(s,a) \leftarrow Q_{t-1}(s,a) + \alpha(R(s,a) + \gamma \max_{a} Q(s',a') - Q_{t-1}(s,a)) \quad \textbf{(6)}$$

where α is the learning rate. The step-by-step working of the algorithm is explained in a study [32].

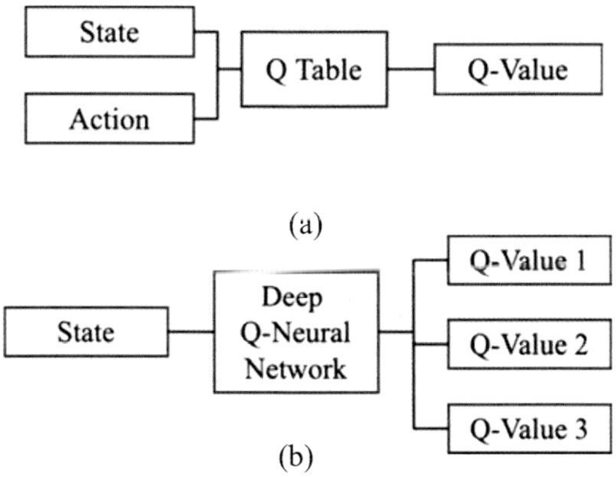

Fig. (4). (a) Q-learning block diagram (b) Deep Q-learning block diagram.

Deep Q-Learning

Q-Learning faces challenges in infinite environments in terms of space complexity. To overcome this challenge, Deep Q-Learning uses Deep Neural Networks to approximate the Q- value function. The input of the neural network used is the parameters that describe the current state and the output of the neural network is Q-values. Functions such as the softmax can then be used to select the best Q-value among the output Q-values. Fig. (**4**) shows the difference between Q-learning and Deep Q-Learning using a block diagram.

In Deep Q-Learning, the weights of the neural network are adjusted according to a squared error loss function, which is expressed as:

$$L = \sum(Q\_Target - Q)^2 \quad \textbf{(7)}$$

Experience Replay

To ensure that the neural network learns a variety of experiences from the environment and does not learn from closely related and contiguous states in an environment, experience replay is implemented. This is important because if closely related states are fed more often in the neural network, the network will learn to perform better only in that particular situation and not throughout the environment. The experiences {s, s′, a, r} are stored in batches in a replay memory, and then random experiences from the batch are selected and fed into the neural network. This ensures a good mix of various kinds of experiences, which helps the neural network adjust better to the complete environment.

SIMULATION EXPERIMENTS

As described above in the article, this section builds on top of the proposed method to showcase the simulation experiments carried out to implement robot path planning.

The Simulator

A simulator was created to simulate the robot path planning environment. The simulator provides the dynamic environment and context necessary for the simulation of the environment in which the path planning of the robot takes place. The entire window of simulation is based on the coordinate system and uses vectors and vector mechanics. The simulator consists of two fixed locations that are defined before the simulation starts, which define the initial position and goal position of the robot. The robot needs to find an optimized path between these two points on the simulator. The simulator comprises some other elements that are described below, along with their uses and properties:

Robot: The robot is shown on the simulator in the form of a rectangle. The robot has 3 sensors attached to it at a distance and varied angles, which are described in detail below. The robot widget has the following attributes attached to it:

- **Orientation:** It is the angle between the direction of the velocity vector and the vector joining the current robot position to the goal. The angle between these two vectors defines the orientation of the robot with respect to the goal.

- **Velocity:** The velocity vector defines the direction and the speed with which the robot moves. Upon hitting an obstacle or moving over an obstacle, the velocity of the robot decreases and is again increased based on the performance of the robot.
- **Angle**: The angle of the robot is simply the angle of the velocity vector with respect to the x-axis. The angle is used by the velocity vector to identify the direction of movement of the robot. The angle of the robot changes based on the prediction of the Deep Q-learning network.

These properties are shown in Fig. (**5**).

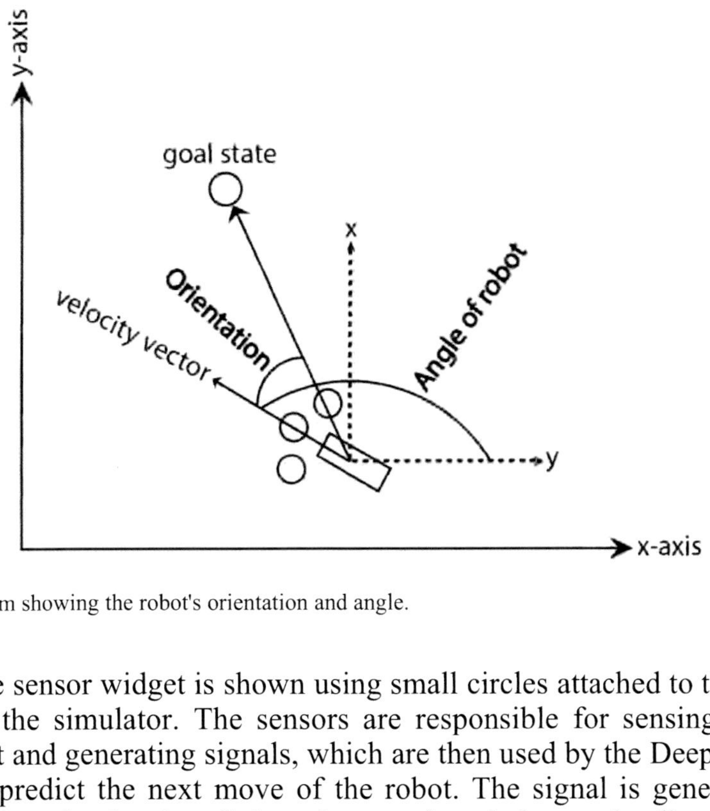

Fig. (5). Diagram showing the robot's orientation and angle.

Sensor: The sensor widget is shown using small circles attached to the robot at a distance in the simulator. The sensors are responsible for sensing the nearby environment and generating signals, which are then used by the Deep Q-Learning network to predict the next move of the robot. The signal is generated by the sensor based on the density of obstacles near it and also on the distance between the obstacle and the sensor. The value of the signal lies between 0 and 1.

Obstacle: The obstacles on the simulator can be drawn in any shape and size and of varying densities in a similar manner as the paint application on Windows. The obstacles can also be drawn during the training phase *i.e.*, obstacles can be introduced at any time. The robot needs to learn to adapt to these obstacles and find an optimized path between the initial state and the goal state.

Buttons: The simulator has 4 buttons which are described below:

- **Clear:** This button is used to clear all the obstacles from the simulator.
- **Save:** This button is used to save the trained Deep Q-Learning model.
- **Load:** This button is used to load a saved Deep Q- Learned model from the file.
- **Start:** This button is used to start the simulation process.

Fig. (**6**) shows an image of the simulator with various elements described above.

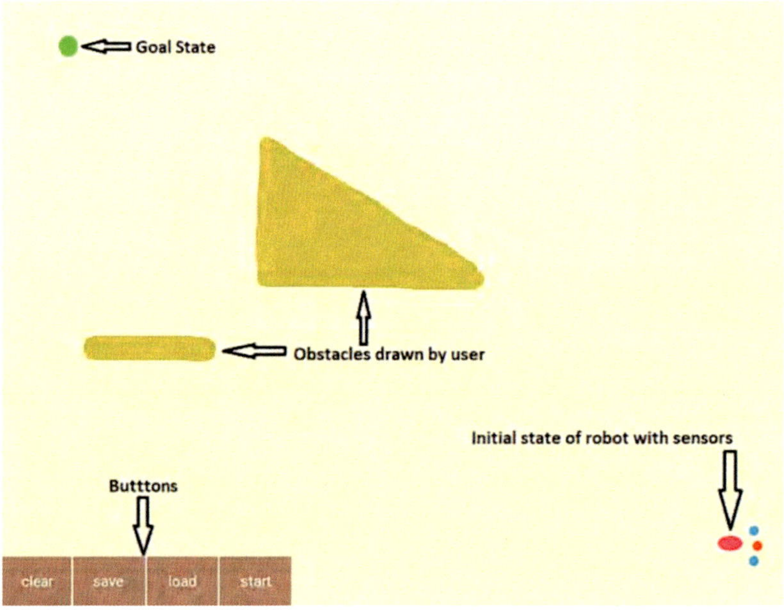

Fig. (6). Description of the simulator.

Deep Q-Learning

The algorithm we use in the implementation to decide the next move of the robot is Deep Q-Learning. Deep Q-Learning is a more advanced version of Q-Learning in which instead of a Q-table, which selects the next move based on the state and the action, a deep neural network is used to predict the Q-values based on certain input parameters of the state. These values are then used to select the action to make the transition to the next state. The Deep Q-Learning neural network architecture we use in simulation is shown in Fig. (**7**). The architecture consists of an input layer, a hidden layer, and an output layer. These layers are explained below in detail.

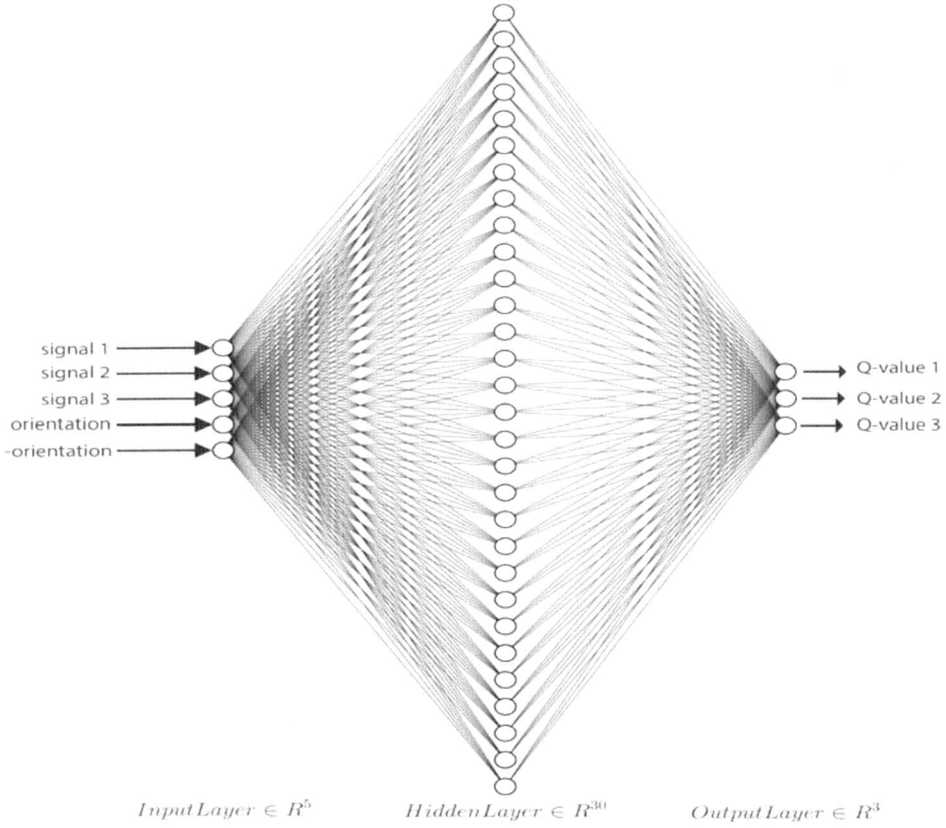

Fig. (7). Deep Q-learning neural network architecture.

The Input layer: The input layer consists of 5 neurons which represent the following 5 parameters:

- **Signal 1:** The first signal is generated from the first sensor of the robot. It measures the distance and the density of the obstacle from the robot on the right-hand side at an angle of 30 degrees from the robot and generates a value that lies between 0 and 1.
- **Signal 2:** The second signal is generated from the second sensor of the robot. It measures the distance and the density of the obstacle in front of the robot and generates a value that lies between 0 and 1.
- **Signal 3:** The third signal is generated from the third sensor of the robot. It measures the distance and the density of the obstacle from the robot on the left-hand side at an angle of 30 degrees from the robot and generates a value that lies between 0 and 1.
- **Orientation:** The orientation parameter of the robot explained above is also

used as an input value and is regularized by dividing the measure of orientation by 180. Hence, this value also lies between -1 and 1.

- **Negative Orientation:** This is just the negation of the orientation input to generate a negative orientation, which is used as the fifth and final input in the input layer of the neural network.

The Hidden Layer: The hidden layer consists of 30 neurons whose values are calculated by using the rectified linear unit activation function. This ensures that the values of hidden layers are always positive. This activation is used as a standard in many Deep Q-Learning models.

The Output Layer: The output Layer consists of 3 output neurons, which generate 3 Q-values. These Q-values are fed into a SoftMax function and converted into probabilities. Based on a multinomial distribution of the probabilities, the action is selected, which rotates the robot and changes the orientation of the robot towards the goal.

Basic Workflow

Fig. (**8**) shows the steps used while working with the simulator. These steps are further elaborated in further subsections:

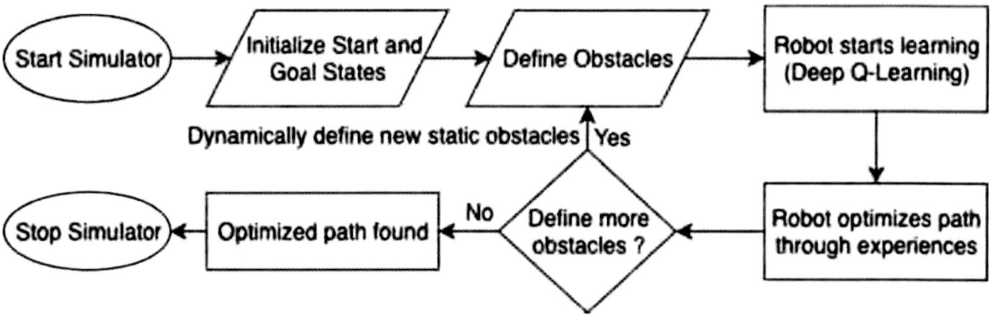

Fig. (8). Basic workflow of the simulator engine.

Definition of Initial and Goal State

Initially, when the simulator is started, the initial state and the goal state are defined by the user and the robot is placed on the initial state. The initial state and the goal state are defined by specifying the x-coordinate and the y-coordinate using the coordinate system. We need to make sure that the initial and the goal states lie within the boundaries of the simulator window. The goal state is indicated by a green dot, and the locus of 100 pixels around the goal state is considered equivalent to reaching to the goal state. Fig. (**9a**) shows the simulator after this step is completed.

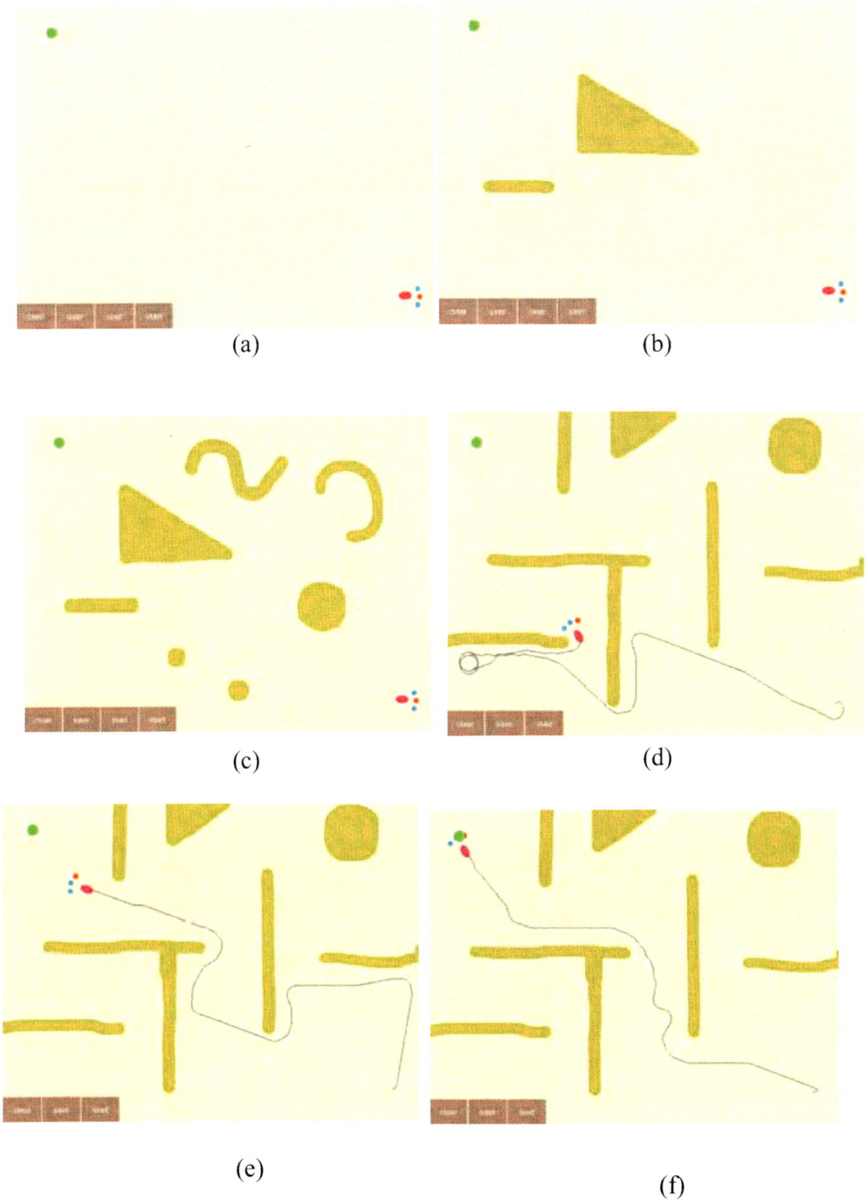

Fig. (9). (a) Initial state of the simulator (b) & (c) Obstacle defintion (c) & (d) Training of robot (e) Path found to goal (f) path found to goal.

Definition of the Obstacles

The obstacles in the simulator can be of any shape and size based on the user's input. To define the obstacles, the user draws on the simulator screen using a cursor similar to the paint application in Windows. The user can also change the density of these obstacles based on the speed with which the user moves the cursor. The density of the obstacle is inversely proportional to the speed of the cursor *i.e.*, the slower the cursor moves, the greater the density of the obstacle and vice versa. Also, once the user is satisfied with the definition of the obstacles, he can press the start button to start the robot training phase. Also, the user can define more obstacles if needed during the training phase by simply drawing on the screen as done during the initial definition of the obstacles. This is shown in Figs. (**9b**) and (**9c**).

Training and Optimization

The robot starts exploring once the start button is pressed and uses the Deep Q-Learning neural network to generate Q-values that are then passed on to a SoftMax function to generate probabilities. Based on these generated values of probabilities, one of the following actions is chosen by the robot to orient itself toward the goal:

(a). Rotate clockwise by 20 degrees.
(b). Continue moving without any rotation.
(c). Rotate anti-clockwise by 20 degrees.

Also, for each change in the state of the robot, the robot is rewarded based on the following rules:

(a). If the robot reaches very near to the simulator boundaries or even hits one of the boundaries, it is rewarded a value of -1.
(b). If the robot hits an obstacle, its velocity is decreased, and it is rewarded a value of -1.
(c). If the robot is in a state from which the distance of the goal is greater than the distance to the goal from the previous state, then the robot is rewarded a value of -0.2.
(d). If the robot is in a state from which the distance of the goal is less than the distance to the goal from the previous state, then the robot is rewarded a value of 0.1.

The velocity of the robot is also decreased if the robot hits or moves over an obstacle, which is then eventually regained to the original velocity if the robot continues to move towards the goal in the right path. After 100 such transitions of state, the batch of these 100 transitions is fed into the neural network to adjust the weights and start the learning process of the neural network. The neural network, on the basis of the actions taken and the rewards received, tries to maximize the rewards. Also, a batch of 100 transitions is fed into the neural network to ensure proper experience replay. This helps the neural network learn according to the complete environment rather than learning with respect to specific areas of the environment. This phase is shown in Figs. (**9d**) and (**9e**).

Optimization of the Path

Eventually, the robot, after many iterations of exploring the environment, finds an optimal path that maximizes the reward earned by the robot. The logic behind the optimal path is the prediction of the right Q-values by the Deep Q-Learning neural network. After the training phase, the weights in the neural network are adjusted by the Adam optimizer in such a way that it starts predicting the right Q-values. Fig. (**9e**) shows an optimal path found by the robot.

Proper robot path planning is done with the help of several artificial intelligence techniques for determining optimum results. Although the techniques are helpful, they cause several issues with respect to design concepts. So, it is necessary to apply cryptography-related security mechanisms. Hence, several post-quantum techniques are used to resolve the related issues [33 - 36]. To apply this quantum technique, there is a need to study or analyze several attack issues and lightweight cryptography techniques [37 - 43]. It helps to design proper secure path planning.

PERFORMANCE EVALUATION

The simulation was carried out under various architectures of the neural network and also under varying parameters. The architecture which gave the best results consisted of 1 hidden layer with 30 nodes. The comparison between various architectures is shown below. The two main plots which were used to compare results among various architectures are described as follows:

Brain Score Plot: This plot shows the score of the robot against the number of decisions taken by the robot. The score of the robot is calculated as the average reward earned by the robot in a thousand decisions. A decision taken by the robot is nothing but a transition from the current state to the next state.

Path Length Plot: This plot shows the distance covered by the robot in each iteration from the initial state to the goal state. This plot helps in determining wether the robot optimizes its path over time or not.

As seen in the graphs shown in Fig. (**5**), some architectures showed results that did not converge into a constant value easily or some that took a long time to converge. Some of the graphs showed a lot of fluctuations and hence were not favorable. Therefore, in our implementation, we finally stuck to the architecture having a single hidden layer with 30 nodes.

Among the various parameters tested, the best results were obtained by the following parameters shown in Table **1**.

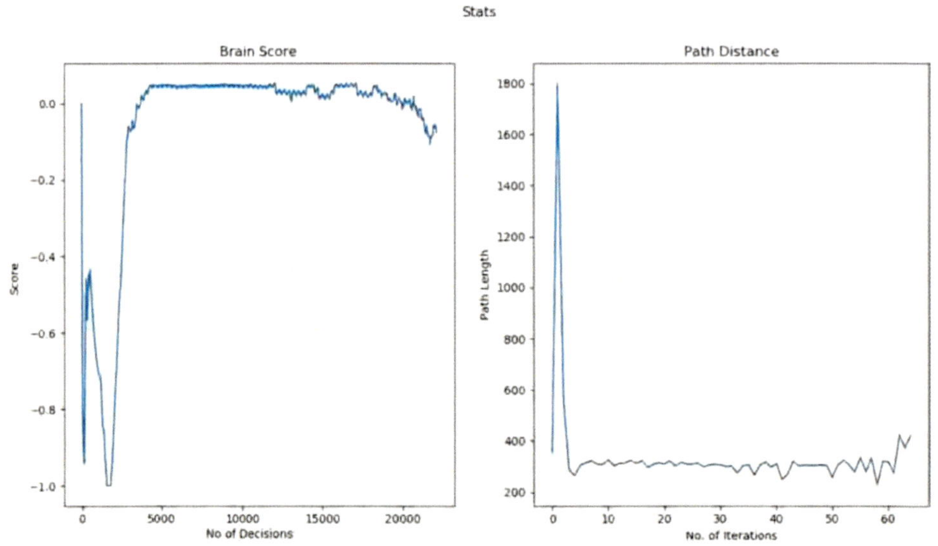

Plot of $R^5 \times R^{15} \times R^3$ Neural Network Architectrue.

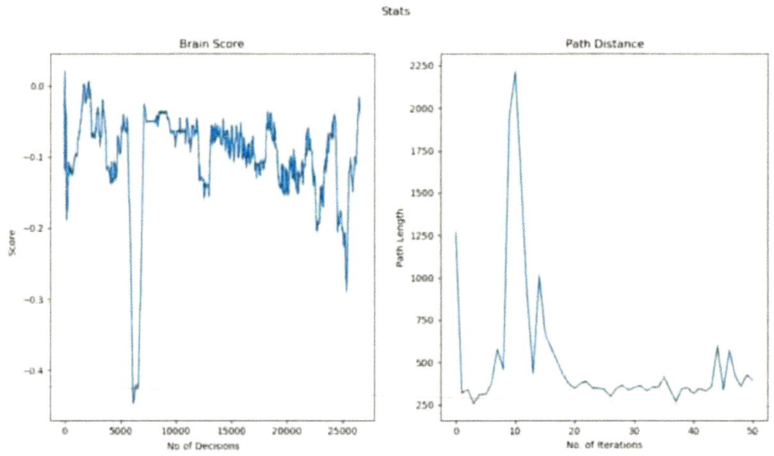

Plot of $R^5 \times R^{15} \times R^{15} \times R^3$ Neural Network Architectrue.

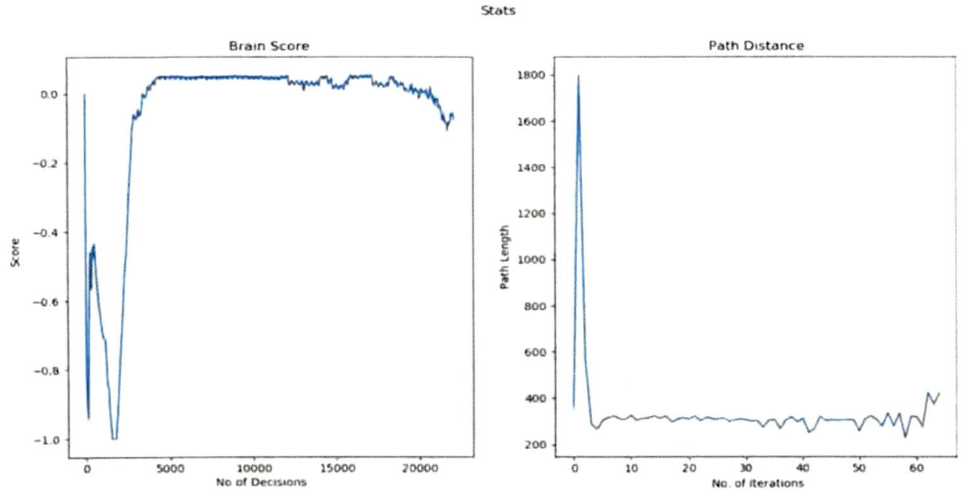

Plot of $R^5 \times R^{15} \times R^{45} \times R^3$ Neural Network Architectrue.

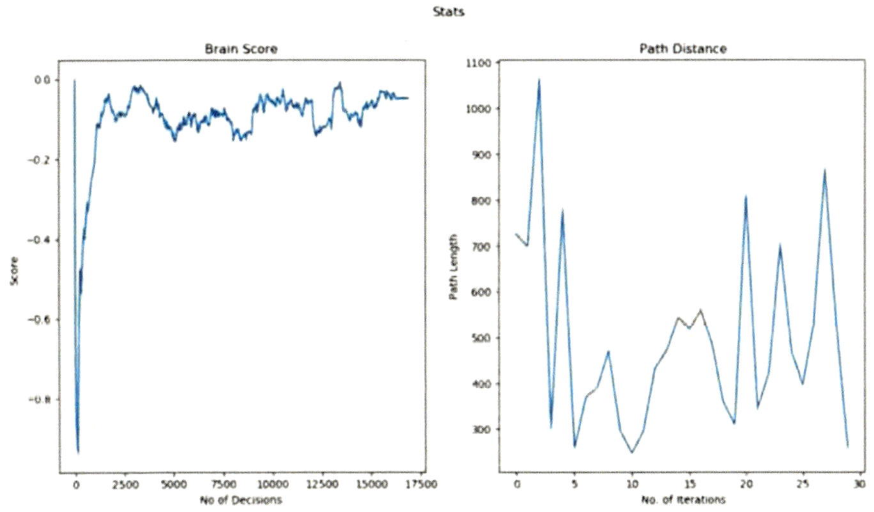

Plot of $R^5 \times R^{30} \times R^3$ Neural Network Architectrue.

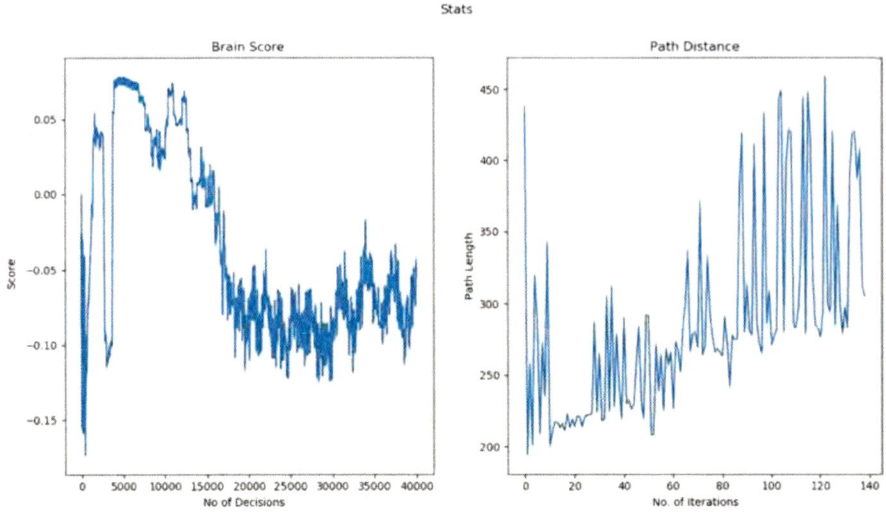

Plot of $R^5 \times R^{30} \times R^{45} \times R^3$ Neural Network Architectrue.

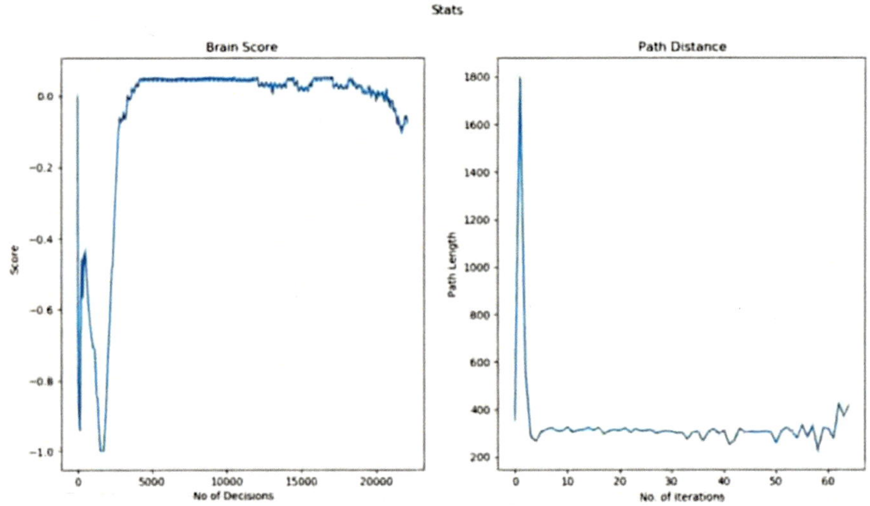

Plot of $R^5 \times R^{45} \times R^3$ Neural Network Architectrue.

Table 1. Table showing various parameters.

Discount Factor (γ)	**0.9**
learning rate (α)	0.001
replay memory size	100000
batch size	100

CONCLUSION

In this paper, we used Deep Q-Learning (Q-Learning using neural networks) to avoid the obstacles in the environment, which were dynamically created by the user. The main aim of the robot is to plan a path without any collision with any of the obstacles. The environment is simulated in the form of a grid that initially contains information on the starting and the target location of the robot. Robots need to plan an obstacle-free path for the given points. The user defines obstacles whenever he/she wishes during the simulation to make the environment dynamic. The accuracy of the path is judged by the path planned by the robot. Various architectures of neural networks are compared, and the simulation results are also analyzed for the evaluation of an optimized path in the dynamic environment.

REFERENCES

[1] C. Chen, Q. Gao, Z. Song, O. Liping, and X. Wu, "Catering service robot", *2010 8th World Congress on Intelligent Control and Automation,* IEEE, pp. 599–604, 2010.

[2] W.P. Chen, S.S. Chen, L.K. Wang, S.P. Cheng, S.C. Chang, and M.F. Lu, "A design of large space automatic cleaning robot with omni-directional wheels", *Catering service robot, in: 2010 8th World Congress on Intelligent Control and Automation,* IEEE, 2010, pp. 599–604, 2013. [http://dx.doi.org/10.1109/iFuzzy.2013.6825425]

[3] P. Das, A.K. Sadhu, R.R. Vyas, A. Konar, and D. Bhattacharyya, "Arduino based multi-robot stick carrying by artificial bee colony optimization algorithm", *Proceedings of the 2015 Third International Conference on Computer, Communication, Control and Information Technology (C3IT),* IEEE, pp. 1–6, 2015. [http://dx.doi.org/10.1109/C3IT.2015.7060152]

[4] W. He, Z. Li, and C. P. Chen, "A survey of human-centered intelligent robots: issues and challenges", *IEEE/CAA Journal of Automatica Sinica,* vol. 4, no. 4, pp. 602-609, 2017.

[5] O. Khatib, O. Brock, K.S. Chang, D. Ruspini, L. Sentis, and S. Viji, "Humancentered robotics and interactive haptic simulation", *Int. J. Robot. Res.,* vol. 23, no. 2, pp. 167-178, 2004. [http://dx.doi.org/10.1177/0278364904041325]

[6] C. Yang, G. Ganesh, S. Haddadin, S. Parusel, A. Albu-Schaeffer, and E. Burdet, "Human-like adaptation of force and impedance in stable and unstable interactions", *IEEE Trans. Robot.,* vol. 27, no. 5, pp. 918-930, 2011. [http://dx.doi.org/10.1109/TRO.2011.2158251]

[7] T. Terzimehic, S. Silajdzic, V. Vajnberger, J. Velagic, and N. Osmic, "Path finding simulator for mobile robot navigation", *XXIII International Symposium on Information, Communication and Automation Technologies,* IEEE, pp. 1-6, 2011. [http://dx.doi.org/10.1109/ICAT.2011.6102086]

[8] P. E. Hart, N. J. Nilsson, and B. Raphael, "A formal basis for the heuristic determination of minimum cost paths", *IEEE Transactions on Systems Science and Cybernetics,* vol. 4, no. 2, pp. 100-107, 1968.

[9] R.E. Korf, "Linear-space best-first search", *Artif. Intell.,* vol. 62, no. 1, pp. 41-78, 1993. [http://dx.doi.org/10.1016/0004-3702(93)90045-D]

[10] M. Likhachev, G.J. Gordon, and S. Thrun, "Anytime a* with provable bounds on sub-optimality", *Advances in neural information processing systems,* pp. 767-774, 2004.

[11] D.D. Harabor, and A. Grastien, "Online graph pruning for pathfinding on grid maps", *Twenty-Fifth AAAI Conference on Artificial Intelligence,* 2011

[http://dx.doi.org/10.1609/aaai.v25i1.7994]

[12] A. Nash, K. Daniel, S. Koenig, and A. Felner, *Theta*: Any-angle path planning on grids.* vol. 7. AAAI, 2007, pp. 1177-1183.

[13] Y. Hu, and S.X. Yang, "A knowledge based genetic algorithm for path planning of a mobile robot", *IEEE International Conference on Robotics and Automation,* vol. 5, pp. 4350-4355, 2004.

[14] A. Tuncer, and M. Yildirim, "Dynamic path planning of mobile robots with improved genetic algorithm", *Comput. Electr. Eng.,* vol. 38, no. 6, pp. 1564-1572, 2012. [http://dx.doi.org/10.1016/j.compeleceng.2012.06.016]

[15] M.A. Contreras-Cruz, V. Ayala-Ramirez, and U.H. Hernandez-Belmonte, "Mobile robot path planning using artificial bee colony and evolutionary programming", *Appl. Soft Comput.,* vol. 30, pp. 319-328, 2015. [http://dx.doi.org/10.1016/j.asoc.2015.01.067]

[16] P.K. Mohanty, and D.R. Parhi, "Optimal path planning for a mobile robot using cuckoo search algorithm", *J. Exp. Theor. Artif. Intell.,* vol. 28, no. 1-2, pp. 35-52, 2016. [http://dx.doi.org/10.1080/0952813X.2014.971442]

[17] B. Patle, A. Pandey, A. Jagadeesh, and D. R. Parhi, "Path planning in uncertain environment by using firefly algorithm", *Defence technology,* vol. 14, no. 6, pp. 691-701, 2018.

[18] L.A. Zadeh, "Fuzzy sets", *Inf. Control,* vol. 8, no. 3, pp. 338-353, 1965. [http://dx.doi.org/10.1016/S0019-9958(65)90241-X]

[19] G. Vachtsevanos, and H. Hexmoor, "A fuzzy logic approach to robotic path planning with obstacle avoidance", *1986 25th IEEE Conference on Decision and Control,* IEEE, pp. 1262–1264, 1986. [http://dx.doi.org/10.1109/CDC.1986.267582]

[20] H. Chang, and T. Jin, "Command fusion based fuzzy controller design for moving obstacle avoidance of mobile robot", In: *Future information communication technology and applications.* Springer, 2013, pp. 905-913. [http://dx.doi.org/10.1007/978-94-007-6516-0_99]

[21] F. Valdez, P. Melin, and O. Castillo, "A survey on nature-inspired optimization algorithms with fuzzy logic for dynamic parameter adaptation", *Expert Syst. Appl.,* vol. 41, no. 14, pp. 6459-6466, 2014. [http://dx.doi.org/10.1016/j.eswa.2014.04.015]

[22] E. Ayari, S. Hadouaj, and K. Ghedira, "A fuzzy logic method for autonomous robot navigation in dynamic and uncertain environment composed with complex traps", In: *Fifth International Multi-conference on Computing in the Global Information Technology.* IEEE, 2010, pp. 18-23. [http://dx.doi.org/10.1109/ICCGI.2010.47]

[23] A. Pandey, R.K. Sonkar, K.K. Pandey, and D. Parhi, "Path planning navigation of mobile robot with obstacles avoidance using fuzzy logic controller", *IEEE 8th International Conference on Intelligent Systems and Control (ISCO),* IEEE, pp. 39-41, 2014. [http://dx.doi.org/10.1109/ISCO.2014.7103914]

[24] L. Cherroun, and M. Boumehraz, "Designing of goal seeking and obstacle avoidance behaviors for a mobile robot using fuzzy techniques", *J. Autom. Syst. Eng.,* vol. 6, no. 4, pp. 164-171, 2012.

[25] M. Zacksenhouse, R.J. DeFigueiredo, and D.H. Johnson, "A neural network architecture for cue-based motion planning", *Proceedings of the 27th IEEE Conference on Decision and Control,* IEEE, pp. 324-327, 1988. [http://dx.doi.org/10.1109/CDC.1988.194321]

[26] E. Nichols, L.J. McDAID, and N.H. Siddique, "Case study on a self-organizing spiking neural network for robot navigation", *Int. J. Neural Syst.,* vol. 20, no. 6, pp. 501-508, 2010. [http://dx.doi.org/10.1142/S0129065710002577] [PMID: 21117272]

[27] E.A. Antonelo, and B. Schrauwen, "Supervised learning of internal models for autonomous goal-

oriented robot navigation using reservoir computing", *2010 IEEE International Conference on Robotics and Automation,* IEEE, pp. 2959-2964, 2010. [http://dx.doi.org/10.1109/ROBOT.2010.5509212]

[28] K-H. Chi, and M-F.R. Lee, "Obstacle avoidance in mobile robot using neural network", *2011 International Conference on Consumer Electronics, Communications and Networks (CECNet),* IEEE, pp. 5082-5085, 2011. [http://dx.doi.org/10.1109/CECNET.2011.5768815]

[29] O. Motlagh, D. Nakhaeinia, S.H. Tang, B. Karasfi, and W. Khaksar, "Automatic navigation of mobile robots in unknown environments", *Neural Comput. Appl.,* vol. 24, no. 7-8, pp. 1569-1581, 2014. [http://dx.doi.org/10.1007/s00521-013-1393-z]

[30] R.S. Pol, and M. Murugan, "A review on indoor human aware autonomous mobile robot navigation through a dynamic environment survey of different path planning algorithm and methods", *2015 International Conference on Industrial Instrumentation and Control (ICIC),* IEEE, pp. 1339-1344, 2015. [http://dx.doi.org/10.1109/IIC.2015.7150956]

[31] R. Bellman, "The theory of dynamic programming", *Bull. Am. Math. Soc.,* vol. 60, no. 6, pp. 503-515, 1954. [http://dx.doi.org/10.1090/S0002-9904-1954-09848-8]

[32] A. Notsu, K. Honda, H. Ichihashi, and Y. Komori, "Simple reinforcement learning for small-memory agent", *2011 10th International Conference on Machine Learning and Applications and Workshops,,* vol. 1, pp. 458-461, 2011. [http://dx.doi.org/10.1109/ICMLA.2011.127]

[33] A. Sarker, "Error detection architectures for hardware/software co-design approaches of number-theoretic transform", *IEEE Trans. Comput. Aided Des. Integrated Circ. Syst.,* 2022.

[34] M.M. Kermani, and R. Azarderakhsh, "Integrating emerging cryptographic engineering research and security education", *2015 ASEE Annual Conference & Exposition,* pp. 26-989, 2015.

[35] M. Mozaffari-Kermani, *Reliable and high-performance hardware architectures for the Advanced Encryption Standard/Galois Counter Mode* Doctoral dissertation, The University of Western Ontario, Canada, 2011.

[36] A. Sarker, M.M. Kermani, and R. Azarderakhsh, "Efficient Error Detection Architectures for Postquantum Signature Falcon's Sampler and KEM SABER. IEEE Transactions on Very Large Scale Integration (VLSI)", *Systems,* vol. 30, no. 6, pp. 794-802, 2022.

[37] A.C. Canto, J. Kaur, M.M. Kermani, and R. Azarderakhsh, "Algorithmic security is insufficient: A comprehensive survey on implementation attacks haunting post-quantum security", *arXiv preprint arXiv:2305.13544,* 2023.

[38] R. Azarderakhsh, B. Koziel, S. F. Langroudi, and M. M. Kermani, "FPGA-SIDH: High-performance implementation of supersingular isogeny Diffie-Hellman key-exchange protocol on FPGA", *Proc. eprint,* pp. 1-18, 2016.

[39] E. Dubrova, K. Ngo, J. Gärtner, and R. Wang, "Breaking a fifth-order masked implementation of crystals-kyber by copy-paste", *Proceedings of the 10th ACM Asia Public-Key Cryptography Workshop,* pp. 10-20, 2023. [http://dx.doi.org/10.1145/3591866.3593072]

[40] J. Kaur, A.C. Canto, M.M. Kermani, and R. Azarderakhsh, "A comprehensive survey on the implementations, attacks, and countermeasures of the current NIST lightweight cryptography standard", *arXiv preprint arXiv:2304.06222,* 2023.

[41] M. M. Kermani, R. Azarderakhsh, and J. Xie, "Error detection reliable architectures of Camellia block cipher applicable to different variants of its substitution boxes", *2016 IEEE Asian Hardware-Oriented Security and Trust (AsianHOST),* pp. 1-6, 2016.

[http://dx.doi.org/10.1109/AsianHOST.2016.7835560]

[42] A. Cintas-Canto, J. Kaur, M. Mozaffari-Kermani, and R. Azarderakhsh, "ChatGPT vs. Lightweight Security: First Work Implementing the NIST Cryptographic Standard ASCON", *arXiv preprint arXiv:2306.08178,* 2023.

[43] M.M. Kermani, E. Savas, and S.J. Upadhyaya, "Guest editorial: Introduction to the special issue on emerging security trends for deeply-embedded computing systems", *IEEE Trans. Emerg. Top. Comput.,* vol. 4, no. 3, pp. 318-320, 2016.
[http://dx.doi.org/10.1109/TETC.2015.2482778]

CHAPTER 3

Seven Concerns of Innovation Design Process and Analysis of a Smart Bicycle Frame

Suraj[1,*], **Arun Kumar**[1] and **Tameshwer Nath**[2]

[1] *National Institute of Technology, Patna, Bihar-800005, India*

[2] *Department of Mechatronics, IIIT Bhagalpur, Bihar-813210, India*

Abstract: Though we are improving our living standards with technology, we are neither bothered about our fitness nor our environment. The article focuses on both, this research will not only help to keep an individual fit but also help in improving air pollution. In this article using the concept of seven concerns of the innovation process, a customized Bicycle frame (Photo Voltaic (PV) frame) is selected after overcoming the flaws in many designs and final conception is analyzed for maximum deformation at different loads. In this article, five concerns out of seven are considered to achieve the final design. Flexible Solar Cell (mentioned in a magazine created by Massachusetts Institute of Technology, Energy Initiative) on the frame will help to charge the battery and that battery will operate other devices like Air Purifier, Mobile Charger, *etc.* that's why this bicycle is called smart bicycle by the author. This article only focuses on innovative design analysis of customized solar frames but not the working of solar frames or air purifiers. This innovative design is going to change the concept of cycling in the future.

Keywords: ANSYS design, Customized bicycle frame, Deformation, Finite element method, Seven concerns of innovation, Solidworks.

INTRODUCTION

The problem associated with air quality has increased with the increase in vehicles running on fossil fuels. As an alternative way of transport, the bicycle is promoted by the Ministry of Housing and Urban Affairs, Government of India. It will help in the reduction in air pollution, energy use, high traffic, noise pollution and allow at the same time a healthier lifestyle for riders. In addition to this, the use of a bicycle is one of the inexpensive and adequate modes of transportation.

Much work has been done on designing and aesthetics of bicycle to improve the quality of riding and increase the acceptance by the individuals as customer has

* **Corresponding author Suraj:** National Institute of Technology, Patna, Bihar-800005, India;
E-mail: suraj.ph21.me@nitp.ac.in

Santosh Kumar Das, Soumi Majumder and Nilanjan Dey (Eds.)

faced very integration uses requirement and customer into design process [1, 2]. As from previous work on frame design and analysis diamond type frame has been found highest rigidity [3]. To manufacture the frame, materials is selected in such a way that it should be light weight, good strength and mostly [4], Just the Design, material and strength of the frame is not enough and so design of fitting system carried some laboratory test and experiments to investigate the aspects of cycling comforts for pilot were also considered in research [5]. Innovative and integrated procedure for optimizing the road bicycle frame, effort was also taken to reduce the deformation of bicycle frame permanently and lessening the mass from the bicycle frame [6]. With the growth of renewable energy and sustainability research is done on power assisted solar bicycle, battery get charged by using of solar energy were solar panel is placed on carriage, solar plate externally placed [7 - 9].

In this article a frame is designed in such a way that the Flexible Solar Cell is on the frame of bicycle. Customized Frame design is finalized after overcoming the flaws in many designs using the con-cept of 7 concerns of innovation. In final customized frame design ANSYS software is used for the structural analysis of the bi-cycle frame and ANSYS software analyze the problems based on the prin-ciple of "Finite Element Method" (FEM).

Concerns of Innovation

There are 7 concerns of innovation, Cause, Context, Comprehen-sion, Check, Conception, Crafting, Connection, from which five concerns of innovation is used for this design.

Cause is that humans nowadays are being affected with fatal disorders like heart syndromes and cancer. The main cause of this is just breathing polluted air of urban cities and not keeping them fit.

Context is that Cycling is one of the most effective ways by which the problem of increasing pollution as well as obesity of an individual can be controlled to some extent.

Comprehension & Check concluded that in today's market there is no such product targeted for sports enthusiast who live in polluted cities and have potential for a very bright future but are un-able to achieve. Therefore, not doing the sports activity like cycling due to the concern of air pollution that may cause irreversible dam-age to their body, this design will solve the problem for those.

Conception gave multiple ideas to address the problems listed in check. Using the Design Process shown in Fig. (**1**) final design was customized.

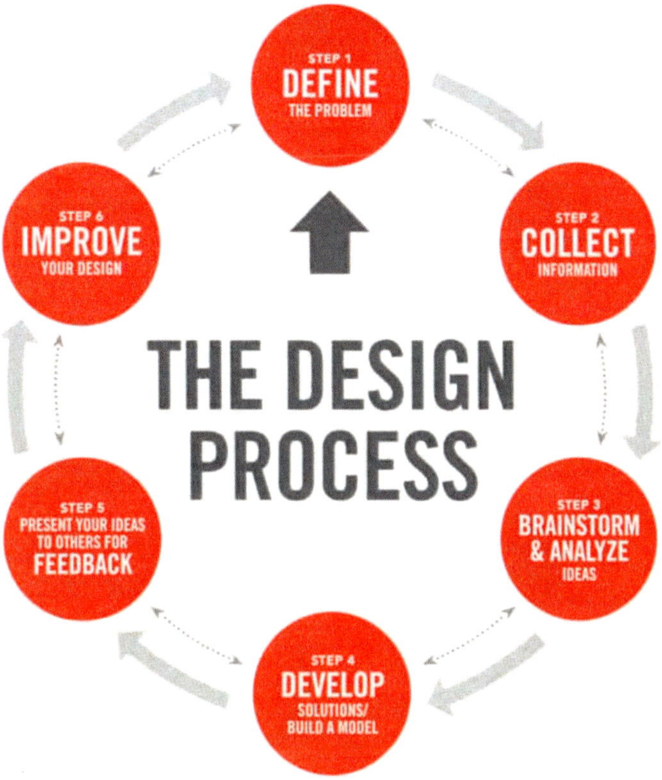

Fig. (1). Design process.

DESIGN CONCEPTS

The main aim of proposed design is to integrate solar panel, bat-tery, and air purifier on to the bi-cycle frame rather than adding them separately. Adding a solar plate separately increases weight due to addition of standoffs and mounting brackets that are needed to hold it in place. This reduces the ergonomics of cycle due to all the wire management issues, reduces ease of maintainability due to all the wires tangling on the frame. It also reduces effectiveness of air puri-fier if it is placed at the back of the bi-cycle frame.

Based on their modification, 4 designs are given:

In Design Concept 1

Main construction: The Bicycle frame, as shown in Fig. (**2a**), incorporates housing for battery and an inverter to charge electron-ics, give power supply to the air purifier and get charged from a solar panel. The solar panel acts as an enclosure to the housing on the bicycle. This enclosure does not need any wires to be separately connected since there are contact pads below the panels that directly connect to the inverter. The solar panel can be easily removed to service or replace the batteries and inverter in case of any problem.

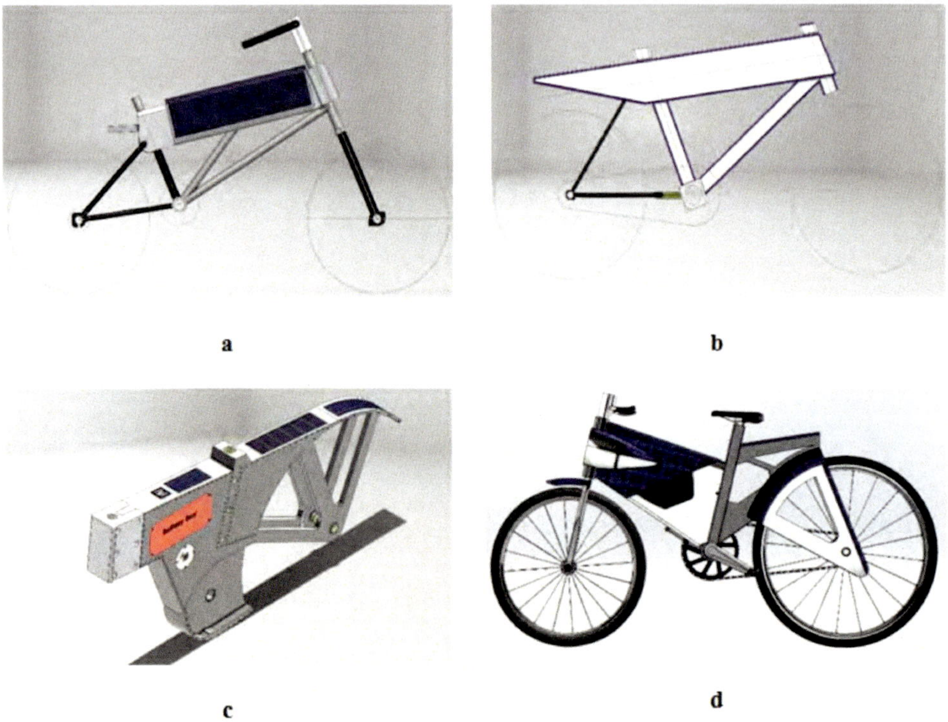

Fig. (2). Design concept phases **a**) Design concept 1, **b**) Design concept 2, **c**) Design concept 3, d) Final design.

Drawbacks: It was not perfect and had some flaws like com-plicated mechanism to clamp a carrier base that would be able to mount the air purifier and the analysis of the cycle showed a heavy deformation of bicycle on the front end of the frame. To overcome flaws in design-1, Design-2 was conceived.

Design Concept 2

Main construction: It overcame some problems faced in de-sign-1 like the front-end frame deformation was reduced and focus was on ergonomics and ease of accessibility and maintainability as shown in Fig. (**2b**). Same solar panel from design-1 was used to enclose the battery and inverter. Other replacement in this design was the clamp mechanism for the carrier was removed and the frame was extended backwards to act as a carrier. This extended carrier would be used to attach all the standoffs and mounting brackets to mount the Air purifier of

$220x600mm^2$. The extended backside frame has an opening that can be used as a storage and to upgrade battery storage if required.

Drawback: The biggest demerit of this design was that it increased the weight of the bicycle and thus making it hard for the cyclist. The other one was the space that could be utilized for storage was extremely limited. To overcome flaws in design-2, Design-3 was conceived.

Design Concept 3

Main Construction: It was meant to be a 2-sided exoskele-ton design that would require forging of a thick sheet for both the sides and then sandwiched with cycle parts like seat, steering handle *etc.* This design as shown in Fig. (**2c**) solved lots of previous prob-lems and included some extra features like a USB outlet to charge your device.

Drawback: However, this design was too bulky and boxy. In addition, in deformation analysis there was a sulk deformation observed where the cyclist would sit. The design was rejected due to high deformation from the middle of the frame. To overcome flaws in design-3, Design-4 was conceived.

Design Concept 4

Main Construction: It is a complete redesign of the other bicycle design variants which are discussed above. This design as shown in Fig. (**2d**) focuses on sleekness, lightness, easy to manufacture and handle. The frame is a bare bone double triangle frame with a sheet metal frame enclosure that encloses an air purifier, battery system and an inverter to charge your electronics for some time. The frame enclosure and front mud guard has integrated flexible solar cells that can charge the batteries and power the air purifier. The air purifier has 3 stages of air filters. The filters can be easily changed by unscrewing the front frame enclosure and the air purifier can be easily accessed. The lower frame enclosure houses battery and in-verter. The air purifier is mounted on the front head of the cycle frame.

One end of hose pipe is connected to purifier and another end is connected to a customized mask, self-designed as shown in Fig. (**3**), which will help cyclist to breathe fresh air while cycling.

Effective: It is designed to be effective in improving health of the cyclist keeping their lungs healthy in polluted cities, efficient in purifying air and elegant to catch the eyes of the buyers. Dimen-sioning of cycle is shown in Fig. (**4**) in 2D.

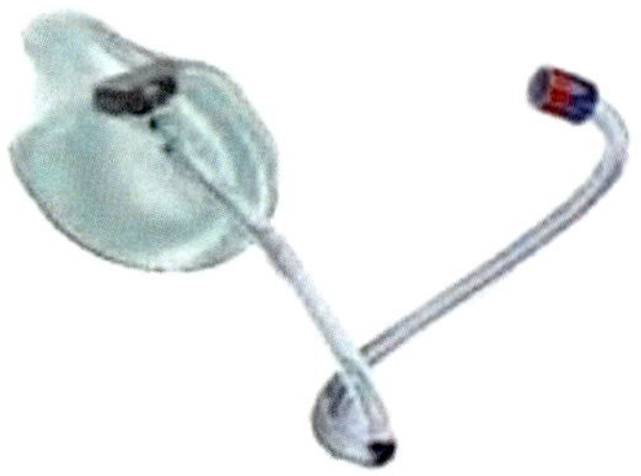

Fig. (3). Customized mask with hose pipe which one end is connected to mask another end.

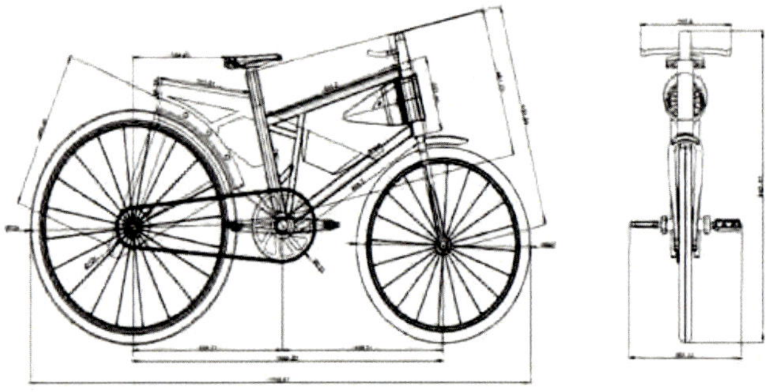

Fig. (4). Dimensioning of cycle is in 2D.

ANALYSIS OF FINAL DESIGNED FRAME

Design made on CAD software as shown in Fig. (**5a**) is trans-ferred to ANSYS, material (Stainless Steel grade 304(SS 304)) is selected and meshing in static structural system is done as shown in Fig. (**5b**).

Stainless Steel grade 304(SS 304) was the material selected instead of other raw materials namely titanium (Ti-64), Aluminium (Al6061-T6) for the analysis of the bicycle frame since:

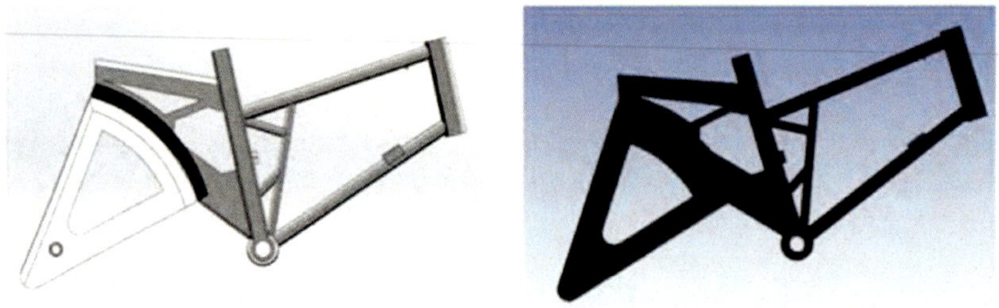

a b

Fig. (5). Frame design and meshing **a**) Design on CAD software **b**) Design Meshing in static structure.

- It is readily available from the market unlike Aluminium and titanium.
- SS 304 has ultimate tensile stress of 505MPa.
- Yield Tensile Strength is around 215Mpa.
- Its elongation at break is 70%.
- Modulus of elasticity is 193Gpa to 200Gpa.
- Has good weldability and machinability characteristics.
- SS 304 costs around Rs.170/kg, whereas Aluminum and titanium costs Rs450/kg and Rs.4000/kg respectively.

In view of factors above, SS 304 was selected for being the material of choice for the frame despite its disadvantage of being heavy. Analysis of deformation is done by giving the boundary conditions and applying different load where the cyclist will be seated. The weights applied on the frame are 800N, 1000N, 1500N and 2000N.

RESULTS AND DISCUSSION

The results obtained for the bicycle frame for the given boundary conditions are as follows, when force of 800N or 80kg was applied on the frame, the total deformation (max) is 0.024 as can be observed in the Fig. (**6a**), Similarly application of force 1250 N or 125kg was applied on the frame, which gave deformation (Max) on the frame was around 0.0386mm, as can be observed from the Fig. (**6b**), When the force of 1500 N or 150 kg was applied on the frame, the total deformation (Max) on the frame was around 0.0463mm, as can be observed from the Fig. (**6c**), When the force of 2000 N or 200 kg was applied on the frame, the total deformation (Max) on the frame was around 0.0617mm, as can be observed from the Fig. (**6d**). It is analyzed that maximum Deformation when maximum load of 2000N is applied is 0.0617mm which can be acceptable. From Table **1** it can be observed that when weight of the cyclist is increased from 80kg to 125kg deformation changes only by 0.0077mm.

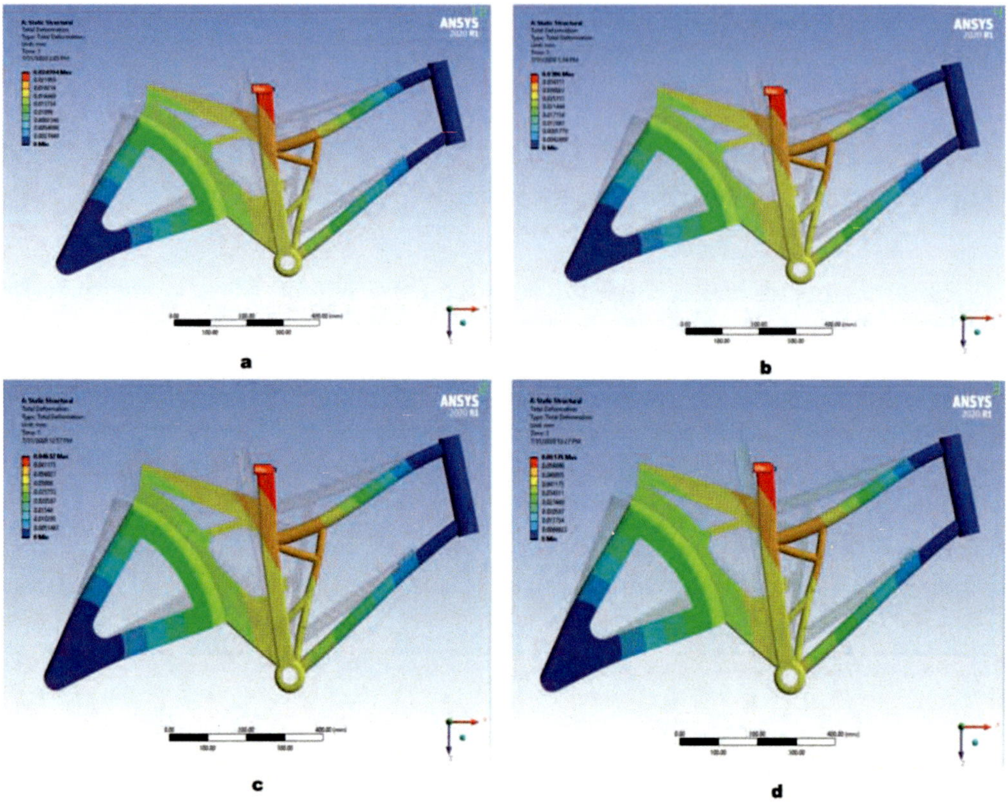

Fig. (6). Deformation analysis at different loads.

Table 1. Deformation in mm at different weight in kg and defor-mation difference.

Body Weight	Deformation of Frame	Deformation Difference
80kg	0.024mm	-
125kg	0.0386mm	0.014mm
150kg	0.0463mm	0.0077mm
200kg	0.0617mm	0.0154mm

CONCLUSION

The material used for the frame of the solar bicycle is Stainless Steel grade 304(SS 304), the frame of the bicycle is rugged. Analysis of the customized frame concludes that the economical weight of the cyclist is 125kg. This design is made only for single person cycling. The final design of the solar framed air purifying cycle is less bulky than the initial designs. This design will be less cumbersome.

The design will be easy to handle by the end user. It will be easy to manufacture, maintain and replace faulty parts. It will also improve the health of the end user by letting him breathe. This Design will change the concept of cycling in the future.

REFERENCES

[1] Z. Siddique, and J.A. Ninan, "Modeling of modularity and scaling for integration of customer in design of engineer-to-order products", *Integr. Comput. Aided Eng.,* vol. 13, no. 2, pp. 133-148, 2006. [http://dx.doi.org/10.3233/ICA-2006-13203]

[2] J.A. Ninan, and Z. Siddique, "Internet-based framework to support integration of the customer in the design of customizable products", *Proc. Inst. Mech. Eng., B J. Eng. Manuf.,* vol. 221, no. 3, pp. 529-538, 2007. [http://dx.doi.org/10.1243/09544054JEM599]

[3] C.C. Lin, S.J. Huang, and C.C. Liu, "Structural analysis and optimization of bicycle frame designs", *Adv. Mech. Eng.,* vol. 9, no. 12, 2017. [http://dx.doi.org/10.1177/1687814017739513]

[4] C. Rontescu, T.D. Cicic, C.G. Amza, O. Chivu, and D. Dobrotă, *Choosing the optimum material for making a bicycle frame,* 2015. https://doaj.org /article/ 7fe5b1f e960 84313 941170460 6c a37d0

[5] H.H.C.M. Christiaans, and A. Bremner, "Comfort on bicycles and the validity of a commercial bicycle fitting system", *Appl. Ergon.,* vol. 29, no. 3, pp. 201-211, 1998. [http://dx.doi.org/10.1016/S0003-6870(97)00052-5] [PMID: 9676337]

[6] Y.C. Cheng, C.K. Lee, and M.T. Tsai, "Multi-objective optimization of an on-road bicycle frame by uniform design and compromise programming", *Adv. Mech. Eng.,* vol. 8, no. 2, 2016. [http://dx.doi.org/10.1177/1687814016632985]

[7] C. Abagnale, M. Cardone, P. Iodice, R. Marialto, S. Strano, M. Terzo, and G. Vorraro, "Design and Development of an Innovative E-Bike", *Energy Procedia,* vol. 101, pp. 774-781, 2016. [http://dx.doi.org/10.1016/j.egypro.2016.11.098]

[8] G. Apostolou, A. Reinders, and K. Geurs, "An Overview of Existing Experiences with Solar-Powered E-Bikes", *Energies,* vol. 11, no. 8, p. 2129, 2018. [http://dx.doi.org/10.3390/en11082129]

[9] K. Mishra, "Design and development of solar hybrid Bicycle", *IJCET,* no. Jan, 2011. [http://dx.doi.org/10.14741/Ijcet/22774106/spl.4.2016.76]

CHAPTER 4

Hybrid Optimization of Profit-Based Unit Commitment Allowing for Uncertainties of Renewable Energy Sources in Summer and Wintertime

Ranadip Roy[1,*], **Ayani Nandi**[1] and **Nirmalya Mallick**[1]

[1] *Department of Electrical Engineering, Sanaka Educational Trust's Group of Institutions, Durgapur, India*

Abstract: Environmental issues, due to the various gases that are harmful to fossil fuels, can cause disease and sickness worldwide. Renewable energy sources (RESs) are a crucial solution for decreasing reliance on fossil fuels. This is because they offer several advantages, such as significant cost reductions in operations, minimal depreciation over time, and the ability to provide electric power for various applications. As a result, they are highly desirable for use in the power sector. This kind of trouble becomes excessively challenging by developing the extent of the electric power market step by step. The authors developed a new optimization technique by combining chaotic maps with various nature-inspired optimization algorithms, such as the Harris hawks optimizer, sine cosine algorithm, and slime mold algorithm. This approach allowed them to improve the performance of these bioinspired optimization methods. The researchers evaluated an improved technique called hCHHO-SCA and hCSMA-SCA for solving the PBUCP considering renewable energy sources. They tested the techniques on both a 10-generating-unit system and a 100-generating-unit system. The authors were able to calculate the profit generated from each system as a result of applying the improved techniques. The adequacy of the analyzer is confirmed for a few benchmark issues that have been observed. The recommended optimizer is helpful in obtaining a solution to problems related to discrete and continuous optimization, including nonlinear types of optimization.

Keywords: Chaotic maps, Harris hawks Optimizer, Metaheuristics, Profit-based unit commitment Problem, Renewable energy sources, Slime mould algorithm, Sine cosine algorithm, Unit commitment problem.

* **Corresponding author Ranadip Roy:** Department of Electrical Engineering, Sanaka Educational Trust's Group of Institutions, Durgapur, India; E-mail: ranadiproy035@gmail.com

Santosh Kumar Das, Soumi Majumder and Nilanjan Dey (Eds.)

INTRODUCTION

India is a developing nation, and each state's progress is influenced by how much electricity it consumes. The need to urgently produce a sufficient amount of electrical power to meet our demand for energy is necessary because the demand for electricity is gradually rising. Advancements in the field of automation have significantly influenced the atmosphere to maintain harmony between power generation and the environment; additionally, the cost of generating electric power with increased reliability is considered by minimizing harm to ecology and the environment, for example, by minimizing the release of flue gases through the use of renewable power sources.

Electrical power systems are expanding rapidly due to the increasing demand for electricity from various sectors, such as commercial, agricultural, residential and industrial sectors. This has made the electrical power sector very large in scale, highly interconnected and complex in nature. The rising electricity demand and changes in the power sector, such as deregulation and privatization, have led to overloading problems in the electric grid. To solve this issue, the electric grid needs to keep pace with the growing electricity demand. Effective power generation scheduling and commitment can help manage the time-varying electrical load demand and optimize the available grid resources [1]. Today's power industry has a variety of sources of electricity, including hydro, thermal, and nuclear power-producing facilities. The demand for electricity varies throughout the day, reaching numerous peaks [2]. Unit commitment is the process of determining which power-generating units should be turned on or off to meet the demand in an effective manner. When deciding to turn units on or off, it is important to consider the cost of starting up or shutting down each generating unit. By analyzing these costs, the optimal generating units can be selected to be turned on when needed in the power system network. The same approach can help decide the order in which generating units should be kept in off condition [3]. The daily demand for power in the commercial, industrial, and residential sectors varies. This approach is similar to how e-mobility fundamentally changes real-time jobs in the power sectors. Thus, it is extremely difficult to meet the fluctuating demand for electricity while incurring the lowest possible production costs.

LITERATURE ANALYSIS

Optimization techniques are an important area of research where scientists are constantly working to improve methods and algorithms. Researchers are developing new optimization strategies to overcome the drawbacks of existing techniques. Many metaheuristic and heuristic optimization methods are currently being used to solve complex problems. Some examples of these methods include genetic algorithms, simulated annealing, tabu search and particle swarm optimization. Overall, optimization research aims to find algorithms that can efficiently find high-quality solutions, as shown in Table **1**.

Table 1. Literature Review of Several Methods.

Glimpse of Some Relevant Work	Reference. No	Name of the Optimizer	Publication's Year
The BEPO method was designed to solve optimization problems with an automatic feature assortment technique. To validate the efficiency of the method, 25 standard benchmarks were used.	[4]	BEPO	2021
The BMRFO technique was implemented to get solutions to feature selection issues. The efficacy of this new method was verified on 18 datasets of the standard functions.	[5]	BMRFO	2021
BSO optimizer had been established to get solutions to the problems associated with optimizing damage detection of structure. It was also used to identify the data related to noise pollution.	[6]	BSO	2021
DSSA technique was designed to get the solution to optimization-related issues. With this method, the exploitation phases were improved to find better optimal points within the search space.	[7]	DSSA	2021
EOA method was implemented to solve the issues in power distributed network reconfigurations and generated electricity allocated in distribution systems.	[8]	EOA	2021
GEPSO optimizer was designed to get solutions for optimization-related problems in local as well as global search space. To verify its efficacy, this method was tested over several well-known standard benchmark functions.	[9]	GEPSO	2021
HHGSO was designed to get solutions for optimization-correlated issues, and this method was applied to improve the phase of exploitations over the search space.	[10]	HHGSO	2021
HLBDA was implemented to get solutions for feature selections in the area of optimization. A case study related to COVID-19 was discussed in this research work.	[11]	HLBDA	2021
A novel IJOA-LV method was designed to get solutions for the existing swarm optimizer, and the viability of this technique was tested over CEC 2011 and CEC 2014 standard functions.	[12]	IJOA-LV	2021
IS-JAYA was used for optimization problems and exploitation and exploration in the search space.	[13]	IS-JAYA	2021

(Table 1) cont.....

Glimpse of Some Relevant Work	Reference. No	Name of the Optimizer	Publication's Year
The MOMBO method was designed to get solutions on cost focused on balancing problems associated with robots.	[14]	MOMBO	2021
The NWOA method was implemented to solve optimization issues, and 25 standard benchmarks were used to test the feasibility of the technique.	[15]	NWOA	2021
The OOBL method was designed to get solutions to engineering optimization problems. Twenty-three benchmarks were used to verify the efficiency of the optimizer.	[16]	OOBL	2021
RPBMO was implemented to get solutions for multilevel feature selection issues.	[17]	RPBMO	2021
The SChoA method was used to get solutions to engineering design optimization problems.	[18]	SChoA	2021
The SHSLTLBO technique was implemented to get solutions to optimization-related problems, and it also improved exploration as well as exploitation in the search space.	[19]	SHSLTLBO	2021
The MFO-SVM method was designed to get solutions to optimization problems related to forecasting the rate of advancement of TBM.	[20]	MFO-SVM	2021
The m-GWO technique was implemented which was suitable to balance exploitation as well as exploration, and was tested over IEEE-CEC-2014 and IEEE-CEC-2017 functions. It was also proved on multilevel thresholding problems in the real world as well as engineering design problems.	[21]	m-GWO	2020
This technique was implemented to find easy solutions to optimization problems, which was verified upon standard IEEE-CEC-2014 benchmarks to determine the efficacy of the method.	[22]	MG SCA	2020
The OAGO method was implemented to get solutions to real-world optimization issues, which was verified over 30 IEEE CEC2017 standard test problems to determine the efficacy of this new method.	[23]	OAGO	2020
A newly made optimizer was designed to get solutions to standard engineering problems, including CEC 2015 benchmarks with statistical analysis and Wilcoxon's test.	[24]	SFOA	2020
This hybrid method was implemented to find solutions to problems with global optimization.	[25]	HC-PSOGWO	2020
ICLBO was developed to find solutions to problems related to engineering design.	[26]	ICLBO	2020

(Table 1) cont.....

Glimpse of Some Relevant Work	Reference. No	Name of the Optimizer	Publication's Year
BMO was implemented to find solutions to problems related to engineering design.	[27]	BMO	2020
The EO method was developed to get the solutions to real-world issues in the optimization area, and it was verified over 58 types of various benchmarks, including uni-modal, multimodal and composite standard functions as well as three kinds of engineering problems.	[28]	EO	2020
The IFDOA technique was developed and verified over the CEC 2019 test system to check its viability to withstand real-world issues.	[29]	IFDOA	2020
The SHO optimizer was one of the most important swarm-based techniques used to get solutions to optimization-related issues by executing its novel exploratory search technique.	[30]	SHO	2020
IWOA was implemented by expending the new tool of the joint search technique to get solutions to optimization problems in global search regions.	[31]	IWOA	2020
MSESCA was designed to get solutions to engineering optimization issues in the real world as well as develop the optimization method in global regions.	[32]	MSESCA	2020
RSHO was implemented and verified to get solutions to various optimization issues at a global level.	[33]	RSHO	2020
IHHO was designed by a combination of SCA as well as HHO methods to solve engineering and numerical issues in the optimization field.	[34]	IHHO	2020
This technique was applied by control-based excellence to get solutions to universal optimization issues.	[35]	GLF–GWO	2019
This proposed technique utilized a cuckoo search algorithm to obtain the best solution.	[42]	CSA	2018

The text discusses various data mining and knowledge management techniques for network and system security [36]. Techniques for detecting system intrusions in networks and progressing information about knowledge management, including knowledge creation, distribution and maintenance, are mentioned [37]. It also discusses smart applications based on user specifications and a nonlinear multiobjective method for hybrid peer-to-peer communications [38]. In summary, the text focuses on different techniques for secure networks, knowledge management and smart applications [39]. The text discusses two research works related to data analytics and machine learning techniques. The first work by Ashish Kumar Das and Santosh Kumar Das proposed using artificial neural networks to optimize data communication in wireless sensor networks [40]. The second work by Nilanjan De discussed data-driven approaches and advances in machine learning and artificial intelligence techniques for healthcare and biomedical applications through data analytics [41].

KEY FEATURES

While current optimization methods have made good progress, research in this area is still at a preliminary stage, with many difficulties yet to be resolved. There are uncertainties around how to calculate a satisfactory state that avoids local or nearby optimal solutions, especially considering the variability of renewable energy sources across seasons. Researchers need to determine how to determine acceptable solutions for both summer and winter conditions when incorporating renewable energy sources into optimization models.

- Overall, additional research is needed to develop optimization methods that can handle the complexities and variability of renewable energy generation.
- Another common operative technique successfully combines the benefits of characteristic enhancement designs with the uncertain nature of RESs in winter and summer.
- First, the full cycle of iterations stops due to certain conditions. However, it does not specify what those conditions are.
- The second main idea is that an important issue is that the process derives brief outcomes within a limited and organized framework using the RES methodology. Overall, the text provides only high-level details about two important aspects of a process without going into specifics.

METHODOLOGIES AND FORMULATION

There are many complex real-world problems related to optimization, machine learning and deep learning that involve constraints and variables. Traditional methods have limitations in solving such complex issues. Metaheuristic methods can handle a large number of optimization problems, but existing research in the field is still at a nascent stage and involves many unresolved issues. There is a need for effective optimization techniques that can help improve solutions and avoid getting stuck at local optima. Every optimizer needs to define parameters such as the number of iterations and cycles within an iteration to execute successfully. Overall, there is scope for improved optimization algorithms that can address the complexities of real-world problems. The text discusses some issues related to short boundaries and hypothetical outlines. The authors propose novel techniques and optimizers to solve these problems in a simple way. The objective is to maximize profit while minimizing total production cost, subject to the system constraints given in the equations. In brief, new techniques for solving these problems are proposed, and the aim is to optimize the profit and cost within the constraints. These terms are given in Eq. (1) to Eq. (3).

$$\text{Profit} = (\text{RV} - TFC) \quad \textbf{(1)}$$

$$\text{RV} = \sum_{h=1}^{H} \sum_{i=1}^{NG} \partial_h \times P_{hi} \times U_{hi} \quad \textbf{(2)}$$

$$\min(TFC) = \sum_{h=1}^{H} \sum_{i=1}^{NG} \{F_{\cosh i}(P_{hi}) + SUC_{hi} + SDC_{hi}\} \quad \textbf{(3)}$$

The overall price of fuel above the scheduled hours 'h' is given in Eq. (4a) and Eq. (4b):

$$TFC = \sum_{h=1}^{H} \sum_{i=1}^{NG} [F_{\cosh i} \times U_{hi} + SUC_{hi}(1 - U_{i,(h-1)}) \times \mathrm{U}_{hi}] \quad \textbf{(4A)}$$

$$TFC = \sum_{h=1}^{H} \sum_{i=1}^{NG} [(A_i P_i^2 + B_i P_i + C_i) \times U_{i,h} + SUC_{i,h}(1 - U_{i,(h-1)}) \times \mathrm{U}_{i,h}] \quad \textbf{(4B)}$$

where the cost of fuel $F_{\cos ih}(P_{ih})$ is calculated in a quadratic manner as convex equations of the power reserve and energy market schedule, which includes the startup cost. The cost of the shutdown of the unit is not considered because it is a very small fraction of the startup costs. In Eq. (5), A_i, B_i and C_i are signified as coefficients of the cost and are represented as $\$/h, \$/MWh$, and $\$/MWh^2$, respectively.

$$F_{\cos ih}(P_i) = A_i P_i^2 + B_i P_i + C_i \quad \textbf{(5)}$$

Cost of the initiating unit

The cost depends on the OFF time, as shown in Eq. (6).

$$SUC_{ih} == \begin{cases} HSU_i; & for \quad T_i^{DW} \le T_i^{UP} \le (T_i^{DW} + T_i^{COLD}) \\ CSU_i; & for \quad T_i^{UP} > (T_i^{DW} + T_i^{COLD}) \end{cases} \quad C_i \qquad (i \in NU \;\; ; \;\; h = 1,2,3,...,H) \quad \textbf{(6)}$$

Portions of these prices are outlined as Eq. (7).

$$SDC_{ih} = KP_{ih} \tag{7}$$

K: additional cost for stopping a unit

Constraint for the power balance to solve the PBUCP with the effect of RES

The mathematical formulation is given in Eq. (8) considering power.

$$\sum_{i=1}^{NG} P_{hi} U_{hi} + P_h^{\mathrm{Re}\,newable} = D_h \tag{8}$$

The present study used hybrid versions of the Chaotic Harris Hawks algorithm combined with the Sine Cosine (hCHHO-SCA) algorithm and the Hybrid Chaotic Slime Mould technique combined with the Sine Cosine optimizer (hCSMA-SCA) to solve the PBUCP of the power sector considering RES in summer and winter to find the optimum scheduling for operational thermal power units as one of the most efficient ways to attain time-varying demands and meet physical and system constraint requirements. To find the hybrid version of the hCHHO-SCA as well as the hCSMA-SCA optimizer, the general operators of the Singer Chaotic Map, Sine Cosine Algorithm, Harris Hawks Optimizer and Slime Mould are utilized. The suggested hybrid optimizers are discussed below in the following sections to obtain solutions to the PBUCP.

HYBRID VERSION OF THE CHAOTIC APPROACH OF HARRIS HAWKS COMBINED WITH THE SINE COSINE OPTIMIZER

In this study, the Harris hawk and sine cosine algorithms are merged with a novel optimization method to create the chaotic Harris hawk optimizer and the Chaotic Sine Cosine Optimizer, which are shown in Fig. (**1**). Some of the key methods used to advance the exploitation phase include explorations involving conversions, exploitation strategies using both hard and soft encircles, soft encircles considering advanced fast dives, and hard encircles considering advanced fast dives. The HHO component is still being discussed. According to the typical pursuit techniques of Harris's hawks, they locate the track and attack their victim using their dominant eyes, making it impossible for the injured person to recognize it. Currently, we can observe analogous options for each of the adjusting systems, which depend on the availability of space for more people to approach in sufficient numbers while being confronted as prey, according to Eq. 9. The implausible nature of society is examined by the Harris's hawks' jump over the track upon their prey, *i.e.*, rabbits. As observed for the rabbits, a surprising jump occurs as a result of the attack, and numerous attacks are achieved *via* the techniques of the era of exploitation eras. These identical chances for every shifting base that takes into account the region for additional people from the transfer to acceptable regions while defying as prey (rabbits) are shown in Eq. (10).

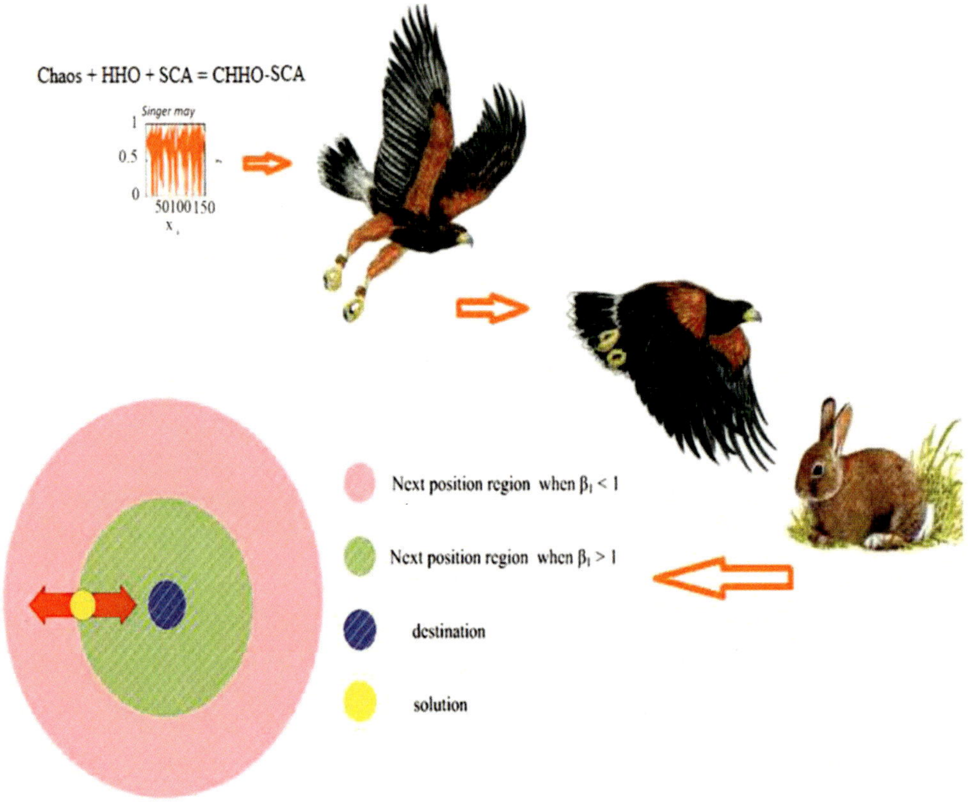

Fig. (1). New hybrid hCHHO-SCA search algorithm.

$$X(\text{iter}+1) = \{X_{RND}(\text{iter}) - r_1 \times abs(X_{RND}(\text{iter}) - 2 \times r_2 \times X(\text{iter})); \quad b \geq .5 \quad \textbf{(9)}$$

$$X(\text{iter}+1) = \{(X_{Rabbit}(\text{iter}) - X_m(\text{iter})) - r_3 \times (L^{\text{limit}} + r_4 \times (U^{\text{limit}} - L^{\text{limit}})); b < .5 \quad \textbf{(10)}$$

$$X_m(\text{iter}) = \frac{1}{P}\left(\sum_{n=1}^{N} X_m(\text{iter})\right) \quad \textbf{(11)}$$

where r_1, r_2, r_3, r_4 and ***b*** are random numbers that are arbitrary in nature in the middle of [0 and 1], which are progressive in nature for every sequential behavior;

X(iter+1) signifies the locations of the rabbit; and *P* signifies the total number of hawks.

The normal hawk ranges are exceeded according to Eq. (11).

Eq. (12) provides a variation after time for the examination of the exploitations;

$$EN = 2 \times EN_0 \times \left(1 - \frac{iter}{iter_{MAX}}\right) \quad \textbf{(12)}$$

HYBRID CHAOTIC SLIME MOULD-SINE COSINE OPTIMIZER

The slime mold optimizer is a newly established stochastic optimizer that involves the oscillation technique of slime molds in nature. Slime molds are referred to as *Physarum polycephalum*. They were first categorized as fungi. This technique frequently simulates further changes in the morphology of slime mold Physarums polycephalum during foraging and does not affect its whole life cycle. Simultaneously, the utilization of weights in the optimizer for imitation of feedback in positive and negative manners produced by slime molds during the search for three diverse morphotypes is a new concept in the research area, as shown in Fig. (**2**).

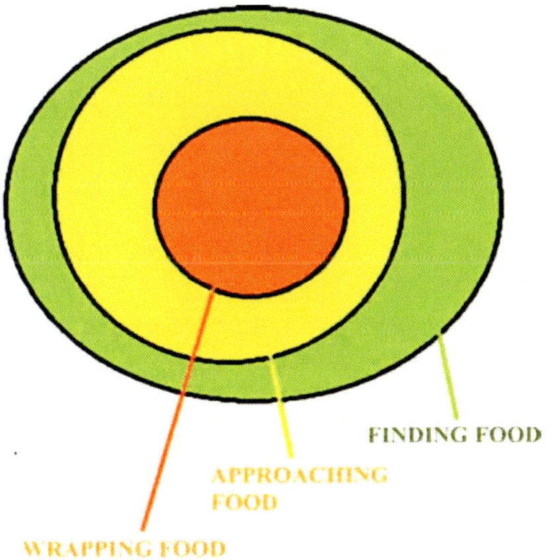

Fig. (2). Basic technique for the SMA optimizer.

SIMULATION ANALYSIS

Ten generating units and several IEEE systems are considered test systems. Hourly profits for the 10th power unit obtained using the hCHHO-SCA optimizer with RES throughout the summer are shown in Table **2**. The energy price and forecasted market price are shown in Table **3**.

Table 2. Profits per hour for the 10th power unit using the hCHHO-SCA optimizer during the summer.

HOURS	PGU1	PGU2	PGU3	PGU4	PGU5	PGU6	PGU7 to PG10	Start-up Cost	Hourly Profit
1	10078.25	5426.75	0	0	0	0	0	0	15505
2	10010	6490	0	0	0	0	0	0	16500
3	10510.5	9124.5	0	0	0	0	0	1680	19635
4	10305.75	10305.75	0	0	0	0	0	560	20611.5
5	10578.75	9648.75	0	3022.5	0	0	0	520	23250
6	10442.25	10442.25	0	2983.5	0	0	0	0	23868
7	10237.5	10237.5	0	2925	0	0	0	0	23400
8	10078.25	9977.4675	0	2879.5	0	0	0	0	22935.2175
9	10374	10374	0	2964	2606.724	0	0	120	26318.724
10	13354.25	13354.25	3815.5	3815.5	4754.7	1193.37101	0	60	40287.571
11	13718.25	13718.25	3919.5	3919.5	4884.3	1291.3245	0	170	41451.1245
12	14400.75	14400.75	4114.5	4114.5	5127.3	1074.201	0	260	43232.001
13	11193	11193	3198	3198	2709.444	0	0	0	31491.444
14	11147.5	11147.5	3185	3185	1850.485	0	0	0	30515.485
15	10237.5	9909.22501	2925	2925	0	0	0	0	25996.725
16	10146.5	9056.922	0	2899	0	0	0	0	22102.422
17	10123.75	8391.81	0	2892.5	0	0	0	60	21408.06
18	10032.75	9291.87	0	2866.5	0	0	0	0	22191.12
19	10101	9666.102	0	2886	0	0	0	0	22653.102
20	10305.75	10158.2985	0	2944.5	0	0	0	0	23408.5485
21	10510.5	10506.804	0	3003	0	0	0	900	24020.304
22	10442.25	10442.25	0	2983.5	0	0	0	0	23868
23	10351.25	10123.75	0	0	0	0	0	0	20475
24	10260.25	7779.75	0	0	0	0	0	0	18040

Table 3. Hourly profits with startup cost and fuel cost for the 40th power unit using the hCHHO-SCA optimizer with RES throughout the summer.

HOURS	Hourly Profit	Cost of Start-up	Cost of Fuel	HOURS	Hourly Profit	Cost of Start-up	Fuel Cost
1	62020	2540	53787.3293	**13**	129793.044	2660	102905.826
2	66000	1980	57283.9693	**14**	126065.485	1800	99701.8277
3	78540	2150	65457.854	**15**	106996.725	120	92283.8666
4	82446	6340	70578.81	**16**	91678.422	900	79499.5794
5	93000	5000	77802.3628	**17**	88158.06	560	75971.4783
6	95472	0	80603.6428	**18**	90987.12	900	79432.2875
7	93600	1120	80603.6428	**19**	91917.102	750	79677.8932
8	92043.2175	0	79941.5826	**20**	94076.5485	750	79907.2172
9	106346.724	10000	90104.7047	**21**	96092.304	290	80018.5558
10	163557.571	5000	107519.601	**22**	95472	10000	80021.3612
11	169166.525	560	108769.819	**23**	81900	770	68711.639
12	177301.401	340	108613.948	**24**	72160	1600	61709.679

The analyses were carried out in the MATLAB 2017a environment. Moreover, the ambient temperature and solar radiation are outlined in Tables **3** and **4**, respectively. The efficiencies of the designed optimizers CHHO, CSCA, CSMA, CHHO-SCA and CSMA-SCA and their outputs are compared, and their practicalities are justified with respect to CHHO, CSCA and CSMA, as discussed in this section. Hourly profits with startup cost and fuel cost for the 40th power unit using the hCHHO-SCA optimizer with RES throughout the summer are shown in Table **3**. Hourly profits with startup and fuel costs for the 100th power unit obtained using the hCHHO-SCA optimizer with RES throughout the summer are shown in Table **4**.

Table 4. Hourly profits with startup and fuel costs for the 100th power unit using the hCHHO-SCA optimizer with RES throughout the summer.

HOURS	Hourly Profit	Cost of Start-up	Cost of Fuel	HOURS	Hourly Profit	Cost of Start-up	Fuel Cost
1	155050	6220	137074.14	**13**	326396.24	470	272503.14
2	165000	10530	147430.92	**14**	317165.49	1720	262129.85
3	196350	4490	165405.74	**15**	268996.73	1850	240643.47
4	206115	3510	176424.66	**16**	230830.42	240	202188.15
5	232500	1090	198131.71	**17**	221658.06	1830	193981.38
6	238680	120	207009.27	**18**	228579.12	4730	205069.3
7	234000	3100	207169.23	**19**	230445.1	860	207679.45

(Table 4) cont.....

HOURS	Hourly Profit	Cost of Start-up	Cost of Fuel	HOURS	Hourly Profit	Cost of Start-up	Fuel Cost
8	230259.22	840	207001.71	**20**	235412.55	4780	209939.67
9	266402.72	11200	235160.62	**21**	240236.3	3370	206357.64
10	410097.57	960	287588.05	**22**	238680	5590	205734.41
11	424597.32	2320	289952.96	**23**	204750	1290	177322.38
12	445440.2	830	290212.87	**24**	180400	7710	158931.94

Hourly profits with startup and fuel costs for the 100th power unit obtained using the hCSMA-SCA optimizer with RES throughout the summer are shown in Table **5**. The statistical results for the PBUCP in the hCHHO-SCA and hCSMA-SCA optimizers with RES throughout the summer are shown in Table **6**. Hypothetical testing of the hCHHO-SCA and hCSMA-SCA optimization algorithms with RES during summer for the PBUCP is shown in Table **7**. A comparison of the computational times for the hCHHO-SCA and hCSMA-SCA optimizers with that of the RES throughout the summer for the PBUCP is shown in Table **8**. A convergence graph of the 10th power unit, 40th power unit and 100th power unit using the hCSMA-SCA optimizer with RES throughout the summer is shown in Fig. (**3**).

Table 5. Hourly profits with startup cost and fuel cost for the 100th power unit using the hCSMA-SCA optimizer with RES throughout the summer.

HOURS	Hourly Profit	Cost of Start-up	Cost of Fuel	HOURS	Hourly Profit	Cost of Start-up	Fuel Cost
1	8240	136089.9752	155050	**13**	700	271913.35	326396.244
2	4950	146507.665	165000	**14**	1830	265131.92	317165.485
3	5450	167890.6885	196350	**15**	1200	243807.251	268996.725
4	3120	187640.593	206115	**16**	3050	208385.842	230830.422
5	2990	207032.5301	232500	**17**	3150	196821.028	221658.06
6	460	213667.8142	238680	**18**	490	210505.037	228579.12
7	1580	212233.989	234000	**19**	2930	207496.206	230445.102
8	2530	210822.389	230259.218	**20**	4240	206778.597	235412.5485
9	16030	237061.118	266402.724	**21**	3700	204094.042	240236.304
10	6210	288693.0248	410097.571	**22**	1870	208967.217	238680
11	1050	292311.407	424597.325	**23**	1540	185358.667	204750
12	380	294438.0778	445440.201	**24**	3820	164275.693	180400

Table 6. Statistical results for the hCSMA-SCA and hCHHO-SCA optimizers with RES throughout the summer.

Test System	10th Power Unit		40th Power Unit		100th Power Unit	
Optimizer	**hCHHO-SCA**	**hCSMA- SCA**	**hCHHO-SCA**	**hCSMA-SCA**	**hCHHO-SCA**	**hCSMA-SCA**
Best Profit	**105592.9791**	92273.6016	**449961.773**	399358.83	**978929.409**	883387.929
Average Profit	105537.0498	92136.57976	414539.117	374868.559	885420.535	832132.856
Worst Profit	104376.4576	91792.55135	377108.946	361952.983	825493.98	805441.246
Std.	231.6005794	133.9500572	17283.7754	8792.59099	38335.0489	17693.7491
Median	105589.6509	92155.3411	409815.915	371648.544	875305.641	827943.407

Table 7. Hypothetical testing of the hCSMA-SCA and hCHHO-SCA optimizers with RES throughout the summer.

Test Systems		10th Power Unit		40th Power Unit		100th Power Unit	
Optimizer		**CHHO- SCA**	**CSMA- SCA**	**CHHO-SCA**	**CSMA-SCA**	**CHHO-SCA**	**CSMA-SCA**
Wilcoxon- rank sum test [p-value]		9.41739E-07	1.63848E-06	1.7344E-06	1.7344E-06	1.7344E-06	1.7344E-06
t-test	7.11522E-79	4.6378E-84	8.4217E-42	4.866E-49	2.5086E-41	2.8316E-50	6.3399E-47
	1	1	1	1	1	1	1

Table 8. Comparison of computational times for the hCSMA-SCA and hCHHO-SCA optimizers with that of RES throughout the summer.

Test System	10th Power Unit		40th Power Unit		100th Power Unit	
Optimizer	**CHHO-SCA**	**CSMA- SCA**	**CHHO- SCA**	**CSMA- SCA**	**CHHO- SCA**	**CSMA- SCA**
Best time (sec.)	0.09375	0.010473	0.4375	0.013527	0.375	0.015625
Average time (sec.)	0.370833333	0.015104167	3.73802083	0.0171875	1.64427083	0.0390625
Worst time (sec.)	0.90625	0.015625	6.828125	0.03125	3.125	0.109375

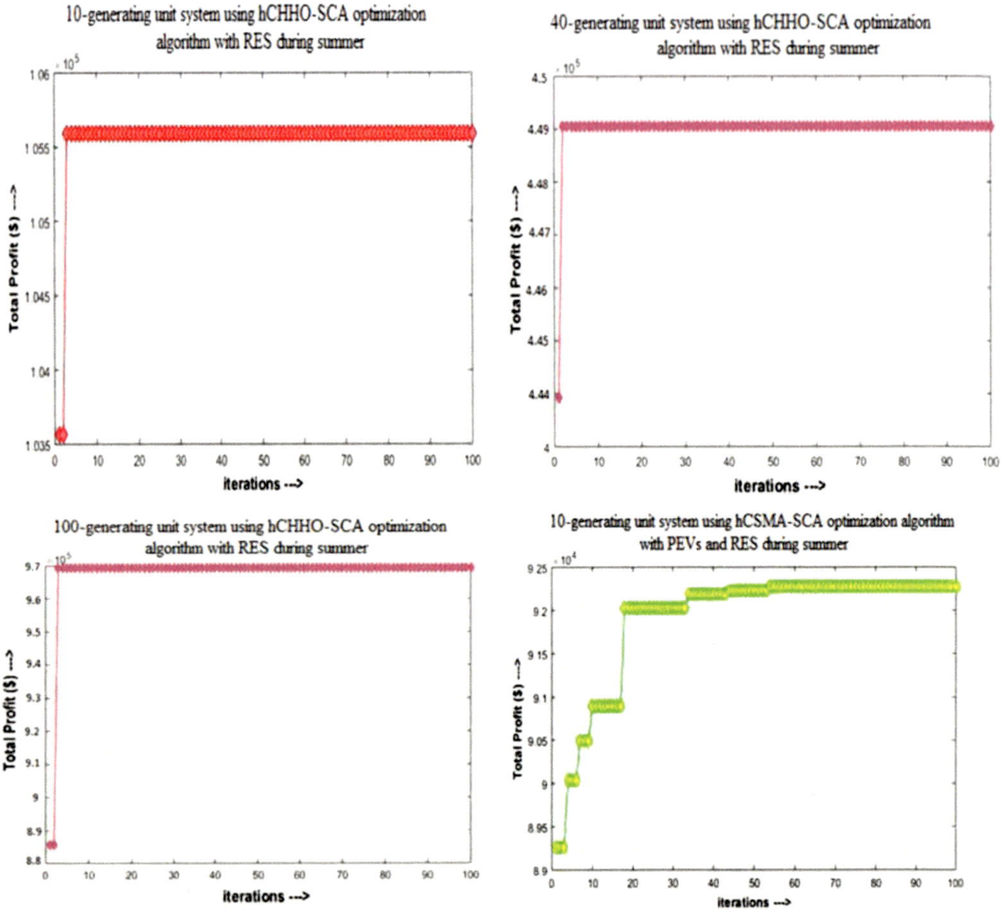

Fig. (3). Convergence graph for the 10th, 40th and 100th power units using the hCSMA-SCA optimizer with RES throughout the summer.

Hourly profit with startup cost and fuel cost for the 100th power unit using the hCSMA-SCA optimizer with RES throughout the winter are shown in Table **9**. The statistical results of the hCSMA-SCA and hCHHO-SCA optimizers with RES throughout the winter for the PBUCP are shown in Table **10**.

Table 9. Hourly profits with startup cost and fuel cost for the 100th power unit using the hCSMA-SCA optimizer with RES throughout the winter.

HOURS	Hourly Profit	Cost of Start-up	Cost of Fuel	HOURS	Hourly Profit	Cost of Start-up	Fuel Cost
1	155050	6820	144861.67	**13**	327131.05	910	272702.94
2	165000	8540	151895.84	**14**	318095.26	580	268026.69

(Table 9) cont.....

HOURS	Hourly Profit	Cost of Start-up	Cost of Fuel	HOURS	Hourly Profit	Cost of Start-up	Fuel Cost
3	196350	2270	169202.5	**15**	269277.3	3950	246217.93
4	206115	2610	180280.99	**16**	231282.22	640	208386.16
5	232500	3490	196212.43	**17**	221986.25	240	197216.44
6	238680	360	209437.7	**18**	228916.04	1110	207299.19
7	234000	3480	210030.66	**19**	230702.84	2750	207051.87
8	230360	1210	208274.37	**20**	235555.92	6150	207073.97
9	266746.94	5940	235888.67	**21**	240240	4050	207213.85
10	410795.51	2310	289460.82	**22**	238680	5070	208001.94
11	425519.01	1110	292817.1	**23**	204750	960	182126.63
12	446327.03	1300	292607.02	**24**	180400	2160	156292.94

Table 10. Statistical analysis of the hCSMA-SCA and hCHHO-SCA optimizers with RES throughout the winter.

Test System	10th Power Unit		40th Power Unit		100th Power Unit	
Optimizer	**hCHHO-SCA**	**hCSMA- SCA**	**hCHHO-SCA**	**hCSMA-SCA**	**hCHHO-SCA**	**hCSMA-SCA**
Best Profit	**106917.3426**	93109.58378	**440563.577**	414311.957	**966825.849**	924140.066
Average Profit	106917.3426	93088.40098	414629.635	378748.854	884044.958	841659.891
Worst Profit	106917.3426	92948.08528	395505.32	366849.122	829806.129	803184.113
Std.	4.44021E-11	48.01551444	11516.7771	12322.1552	36595.9704	28996.8621
Median	106917.3426	93109.58378	414595.288	375010.293	871951.01	836654.764

Hypothetical testing of the hCSMA-SCA and hCHHO-SCA optimizers with RES throughout the winter for the PBUCP is shown in Table **11**. A comparison of the computational times of the hCSMA-SCA and hCHHO-SCA optimizers considering RESs during winter is shown in Table **12**. Table **13** represents the proposed optimizers.

Table 11. Hypothetical testing of the hCSMA-SCA and hCHHO-SCA optimizers with RES throughout the winter.

Test System		10th Power Unit		40th Power Unit		100th Power Unit	
Optimizer		hCHHO-SCA	hCSMA-SCA	hCHHO-SCA	hCSMA-SCA	hCHHO-SCA	hCSMA-SCA
Wilcoxon-rank sum test [p-value]		4.32046E-08	4.16235E-07	1.7344E-06	1.7344E-06	1.7344E-06	1.7344E-06
t- test	0	4.12671E-97	6.5376E-47	6.3827E-45	6.8428E-42	3.3554E-44	1.2926E-46
	1	1	1	1	1	1	1

Table 12. Comparison of computational times for the hCSMA-SCA and hCHHO-SCA optimizers with RES throughout the winter.

Test System	10th Power Unit		40th Power Unit		100th Power Unit	
Optimizer	hCHHO-SCA	hCSMA- SCA	hCHHO-SCA	hCSMA-SCA	hCHHO-SCA	hCSMA-SCA
Best time (sec.)	0.109375	0.014374	0.203125	0.015625	0.40625	0.017851
Average time (sec.)	0.340625	0.015104167	1.30520833	0.01614583	1.71927083	0.04010417
Worst time (sec.)	0.765625	0.015625	3.953125	0.03125	3.21875	0.109375

Table 13. Abbreviation of the Proposed Optimizers.

SI No.	Abbriviation of proposed optimizer	Full Name of proposed optimizer
1	hCHHO-SCA	Hybrid Version of the Chaotic Approach of Harris Hawks combine with Sine Cosine Optimizer
2	hCSMA- SCA	Hybrid Chaotic Slime Mould-Sine Cosine Optimizer

CONCLUSION

One of the difficult problems in power system operation control and planning, the PBUCP, was mathematically formulated in the suggested study effort considering battery electric vehicles, plug-in electric vehicles, and renewable energy sources (solar power). Researchers working on these types of problems with renewable energy sources (RESs) as one of their research aims will benefit from the numerical construction of the PBUCP that has been provided. The 10-, 40-, and 100-generating unit test systems have been effectively scheduled to maximize revenues utilizing proposed hybrid optimizers. The performance of the hCHHO-SCA optimizer is compared to that of other existing heuristic, metaheuristic, and evolutionary search optimizers, and it is determined that the performance of the hCHHO-SCA is superior. The hCHHO-SCA and hCSMA-SCA optimization techniques, as well as the CHHO and CSCA methods for solving the PBUCP problem while impacting the RES on summer and winter days, were successfully demonstrated. The performance of the suggested hybrid optimized method for a standard test system, which consists of thermal generating units for small-, medium-, and large-scale power sectors, was also evaluated. The effectiveness of the proposed method was evaluated for systems with 10 generating units, 40 generating units, and 100 generating units. After a successful trial, it was found that the recommended optimizer is superior to other current optimizers in terms of its ability to solve continuous, discrete, and nonlinear optimization problems.

REFERENCES

[1] R.H. Kerr, "Unit commitment", *EEE Trans. Power App. Sys.,* vol. 85, no. 5, pp. 417-421, 1966.

[2] C.J. Baldwin, K.M. Dale, R.F. Dittrich, and A. Study, "Study of economic shutdown of generating units in daily dispatch, AIEE transaction of power apparatus and systems", *Vol. PAS,* vol. 78, pp. 1272-1284, 1959.

[3] K.D. Lee, R.H. Vierra, G.D. Nagel, R.T. Jenkins, vol. 104, no. 8, pp. 2072-2078, 1985.

[4] S. Vimal, A. Sharma, and K. Cengiz, "BEPO: A novel binary emperor penguin optimizer for automatic feature selection", *Knowl.-Based Syst.,* vol. 211, p. 106560, 2020. [http://dx.doi.org/10.1016/j.knosys.2020.106560]

[5] K.K. Ghosh, R. Guha, S.K. Bera, N. Kumar, and R. Sarkar, "S-shaped versus V-shaped transfer functions for binary Manta ray foraging optimization in feature selection problem", *Neural Comput Appl.,* vol. 33, no. 17, pp. 11027-11041, 2021. [http://dx.doi.org/10.1007/s00521-020-05560-9]

[6] A. Kaveh, S.M. Hosseini, and A. Zaerreza, "Boundary strategy for optimization-based structural damage detection problem using metaheuristic algorithms", *Period. Polytech.: Civ. Eng.,* vol. 65, no. 1, pp. 150-167, 2021.

[7] M. Tubishat, "Dynamic Salp swarm algorithm for feature selection", *Expert Syst Appl.,* vol. 164, p. 113873, 2021. [http://dx.doi.org/10.1016/j.eswa.2020.113873]

[8] A.M. Shaheen, A.M. Elsayed, R.A. El-sehiemy, and A.Y. Abdelaziz, "Equilibrium optimization algorithm for network reconfiguration and distributed generation allocation in power systems", *Appl. Soft Comput.,* vol. 98, p. 106867, 2020. [http://dx.doi.org/10.1016/j.asoc.2020.106867]

[9] D. Sedighizadeh, E. Masehian, M. Sedighizadeh, and H. Akbaripour, "GEPSO: A new generalized particle swarm optimization algorithm", *Math. Comput. Simul.,* vol. 179, pp. 194-212, 2021. [http://dx.doi.org/10.1016/j.matcom.2020.08.013]

[10] K.Z. Zamli, M.A. Kader, S. Azad, and B.S. Ahmed, "Hybrid Henry gas solubility optimization algorithm with dynamic cluster-to-algorithm mapping", *Neural Comput. Appl.,* vol. 33, no. 14, pp. 8389-8416, 2021.

[http://dx.doi.org/10.1007/s00521-020-05594-z]

[11] J. Too, and S. Mirjalili, "A hyper learning binary dragonfly algorithm for feature selection: A COVID-19 case study", *Knowledge-Bas. Syst.,* vol. 212, p. 106553, 2020. [http://dx.doi.org/10.1016/j.knosys.2020.106553]

[12] G. Iacca, V. Celso, V. Veloso, and D. Melo, "An improved Jaya optimization algorithm with Lévy flight", *Expert Syst Appl.,* vol. 165, p. 113902, 2021. [http://dx.doi.org/10.1016/j.eswa.2020.113902]

[13] A. Kaveh, S.M. Hosseini, and A. Zaerreza, "Improved Shuffled Jaya algorithm for sizing optimization of skeletal structures with discrete variables", *Structures,* vol. 29, pp. 107-128, 2021. [http://dx.doi.org/10.1016/j.istruc.2020.11.008]

[14] Z. Li, M.N. Janardhanan, and Q. Tang, "Multi-objective migrating bird optimization algorithm for cost-oriented assembly line balancing problem with collaborative robots", *Neural Comput. Appl.,* vol. 33, no. 14, pp. 8575-8596, 2021. [http://dx.doi.org/10.1007/s00521-020-05610-2]

[15] A.M. Basset, R. Mohamed, and S. Mirjalili, "A novel whale optimization algorithm integrated with nelder–mead simplex for multi-objective optimization problems", *Knowl.-Based Syst.,* vol. 212, p. 106619, 2020. [http://dx.doi.org/10.1016/j.knosys.2020.106619]

[16] W. Lei, X. Kwok, and Y. Zhao, *An orthogonal opposition based learning Yin Yang pair optimization algorithm for engineering optimization, no. 0123456789* Springer: London, 2021. [http://dx.doi.org/10.1007/s00366-020-01248-9]

[17] A.A. Bidgoli, E.H. Komleh, and S. Rahnamayan, "Reference-point-based multi-objective optimization algorithm with opposition-based voting scheme for multi-label feature selection", *Inf. Sci.,* vol. 547, pp. 1-17, 2021. [http://dx.doi.org/10.1016/j.ins.2020.08.004]

[18] M. Kaur, R. Kaur, N. Singh, and G. Dhiman, "SChoA : A newly fusion of sine and cosine with chimp optimization algorithm for HLS of datapaths in digital filters and engineering applications", *Eng. Comput.,* vol. 38, no. S2, pp. 975-1003, 2021. [http://dx.doi.org/10.1007/s00366-020-01233-2]

[19] Z. Chen, Y. Liu, Z. Yang, X. Fu, J. Tan, and X. Yang, "An enhanced teaching-learning-based optimization algorithm with self-adaptive and learning operators and its search bias towards origin", *Swarm Evol. Comput.,* vol. 60, p. 100766, 2021. [http://dx.doi.org/10.1016/j.swevo.2020.100766]

[20] J. Zhou, Y. Qiu, S. Zhu, D. Jahed, and C. Li, "Optimization of support vector machine through the use of metaheuristic algorithms in forecasting TBM advance rate", *Eng. Appl. Artif. Intell.,* vol. 97, p. 104015, 2021. [http://dx.doi.org/10.1016/j.engappai.2020.104015]

[21] S. Gupta, and K. Deep, "A memory-based Grey Wolf Optimizer for global optimization tasks", *Appl. Soft Comput.,* vol. 93, p. 106367, 2020. [http://dx.doi.org/10.1016/j.asoc.2020.106367]

[22] S. Gupta, K. Deep, and A.P. Engelbrecht, "A memory guided sine cosine algorithm for global optimization", *Eng. Appl. Artif. Intell.,* vol. 93, no. 2-4, p. 103718, 2020. [http://dx.doi.org/10.1016/j.engappai.2020.103718]

[23] Z. Xu, "Orthogonally-designed adapted grasshopper optimization: A comprehensive analysis", *Expert Syst Appl.,* vol. 150, p. 113282, 2020. [http://dx.doi.org/10.1016/j.eswa.2020.113282]

[24] D. Sattar, and R. Salim, "A smart metaheuristic algorithm for solving engineering problems", *Eng. Comput.,* no. 0123456789, 2020.

[http://dx.doi.org/10.1007/s00366-020-00951-x]

[25] N. Banerjee, "HC-PSOGWO : Hybrid crossover oriented PSO and GWO based co-evolution for global optimization", *2019 IEEE Region 10 Symposium (TENSYMP),* 2019 07-09 June 2019, Kolkata, India [http://dx.doi.org/10.1109/TENSYMP46218.2019.8971231]

[26] M. Shahrouzi, and A. Salehi, "Imperialist competitive learner-based optimization: A hybrid method to solve engineering problems", *Int. J. Opt. Civ. Eng.,* vol. 10, no. 1, pp. 155-180, 2020.

[27] M. Herwan, Z. Mustaffa, M. Mawardi, and H. Daniyal, "Barnacles mating optimizer: A new bio-inspired algorithm for solving engineering optimization problems", *Eng. Appl. Artif. Intell.,* vol. 87, p. 103330, 2020. [http://dx.doi.org/10.1016/j.engappai.2019.103330]

[28] A. Faramarzi, M. Heidarinejad, B. Stephens, and S. Mirjalili, "Equilibrium optimizer: A novel optimization algorithm", *Knowled-Bas. Syst.,* vol. 191, p. 105190, 2019. [http://dx.doi.org/10.1016/j.knosys.2019.105190]

[29] D.A. Muhammed, S.A.M. Saeed, T.A. Rashid, and I. Member, "Improved algorithm fitness : Dependent optimizer", *IEEE Access,* vol. 8, pp. 19074-19088, 2020. [http://dx.doi.org/10.1109/ACCESS.2020.2968064]

[30] N. Panda, "Improved spotted hyena optimizer with space transformational search for training pi-sigma higher order neural network", *Computat. Intelligen.,* vol. 40, no. 1, pp. 320-350, 2019. [http://dx.doi.org/10.1111/coin.12272]

[31] Q. Fan, Z. Chen, Z. Li, Z. Xia, J. Yu, and D. Wang, "A new improved whale optimization algorithm with joint search mechanisms for high dimensional global optimization problems", *Eng. Comput.,* no. 0123456789, 2020. [http://dx.doi.org/10.1007/s00366-019-00917-8]

[32] H. Chen, M. Wang, and X. Zhao, "A multi-strategy enhanced sine cosine algorithm for global optimization and constrained practical engineering problems", *Appl. Math. Comput.,* vol. 369, p. 124872, 2020. [http://dx.doi.org/10.1016/j.amc.2019.124872]

[33] A. Yimit, K. Iigura, and Y. Hagihara, "Refined selfish herd optimizer for global optimization problems", *Expert Syst. Appl.,* vol. 139, p. 112838, 2020. [http://dx.doi.org/10.1016/j.eswa.2019.112838]

[34] V.K. Kamboj, A. Nandi, A. Bhadoria, and S. Sehgal, "An intensify Harris Hawks optimizer for numerical and engineering optimization problems", *Appl. Soft Comput. J.,* vol. 89, p. 106018, 2019. [http://dx.doi.org/10.1016/j.asoc.2019.106018]

[35] S. Gupta, and K. Deep, "Enhanced leadership inspired grey wolf optimizer for global optimization problems", *Eng. Comput.,* no. 0123456789, 2019. [http://dx.doi.org/10.1007/s00366-019-00795-0]

[36] S.K. Das, and N. Dey, *Constraint Decision-Making Systems in Engineering.* IGI Global, 2023. [http://dx.doi.org/10.4018/978-1-6684-7343-6]

[37] M. Soumi, and D. Nilanjan, *AI-empowered Knowledge Management* Springer, 2022. [http://dx.doi.org/10.1007/978-981-19-0316-8]

[38] S. Das, "Smart design and its applications: Challenges and techniques", In: *Nature-Inspired Computing for Smart Application Design*, 2021, pp. 1-6. [http://dx.doi.org/10.1007/978-981-33-6195-9_1]

[39] S. Das, N. Dey, R.G. Crespo, and H.E. Viedma, "A non-linear multi-objective technique for hybrid peer-to-peer communication", *Informat. Sci.,* vol. 629, pp. 413-439, 2023. [http://dx.doi.org/10.1016/j.ins.2023.01.117]

[40] A.K. Das, and S.K. Das, "An optimal data communication system in wireless sensor network using

artificial neural network", In: *Constraint Decision-Making Systems in Engineering* IGI-GLOBAL pp. 120-134.
[http://dx.doi.org/10.4018/978-1-6684-7343-6.ch007]

[41] N. De, *Data-Driven Approach for Biomedical and Healthcare.* Springer Nature, 2022.

[42] S. Maghsudlu, and S. Mohammadi, "Optimal scheduled unit commitment considering suitable power of electric vehicle and photovoltaic uncertainty", *J. Renewable Sustain. Energy,* vol. 10, p. 043705, 2018.
[http://dx.doi.org/10.1063/1.5009247]

CHAPTER 5

The Impact of Industry 4.0 on Manufacturing: Challenges and Opportunities

Ramesh Chandra Goswami[1], **Hiren Joshi**[1] and **Sunil Gautam**[2,*]

[1] *Gujarat University Ahmedabad, Gujarat, India*

[2] *Department of Computer Science and Engineering, Institute of Technology, Nirma University, Ahmedabad, Gujarat-382481, India*

Abstract: The concept of Industry 4.0 emerged in Europe a decade ago, and later, it was investigated and adopted by academics and industries throughout the world. Due to significant technical improvements in various industries over the past few years, the world's industrial systems have transformed. The major technologies, such as IoT and Big Data, have a large impact on Industry 4.0. They affect each and every sector of the economy. The main advantages include advances in productivity, efficiency, flexibility, decision-making process, and quality of goods and services. The challenges include analyzing the data produced and integrating new technology with the staff and equipment that are already in place. Our goal is to summarize the potential and challenges associated with adopting Industry 4.0 in the manufacturing sector.

Keywords: Digital transformation, Industry 4.0, Manufacturing companies, Smart manufacturing, Smart factory, Smart supply.

INTRODUCTION

Industry 4.0 is a ground-breaking method of organizing industrial production that blurs the lines between different physical objects and unifies them into one complex. It is based on a system of interdependent and connected industrial components that has undergone considerable digitization and automation of production and distribution operations [1]. Industry 4.0 is transforming how businesses produce, enhance, and distribute their goods. Manufacturers are incorporating new technology into their operations and manufacturing facilities, such as the Internet of Things (IoT), cloud computing and analytics, AI, and machine learning.

* **Corresponding author Sunil Gautam:** Department of Computer Science and Engineering, Institute of Technology, Nirma University, Ahmedabad, Gujarat-382481, India; E-mail: sunil.gautam@nirmauni.ac.in

Advanced sensors, embedded software, and robotics are all featured in these "smart factories", which collect and analyze data to help with decision-making. When operational data from ERP, supply chain, customer service, and othercorporate systems is linked with data from production operations, even more value is generated from the previously segregated information [2]. Increased automation, preventative maintenance, self-optimization of process improvements, and most importantly, a new level of efficiency and customer responsiveness made possible by digital technology are all results of their use [3].

The manufacturing sector has a fantastic potential to join the Fourth Industrial Revolution by developing smart factories [4]. Real-time visibility of manufacturing assets is ensured by analyzing the massive amounts of big data gathered from sensors on the factory floor, which can also provide tools for carrying out predictive maintenance to reduce equipment downtime [5].

Smart factories with IoT technology have higher production and better quality. Manufacturing errors are decreased, and money and time are saved when manual inspection business models are replaced with AI-powered visual insights [6]. A smartphone connected to the cloud can be easily set up by quality control workers to enable remote monitoring of manufacturing processes. Manufacturers can identify mistakes earlier rather than later, when repair work is more expensive, by using machine learning algorithms [7].

All types of industrial businesses, including discrete and process manufacturing, as well as oil and gas, mining, and other industrial areas, can use Industry 4.0 concepts and technologies [8]. Fig. (**1**) shows the transformation from smart factory to smart manufacturing.

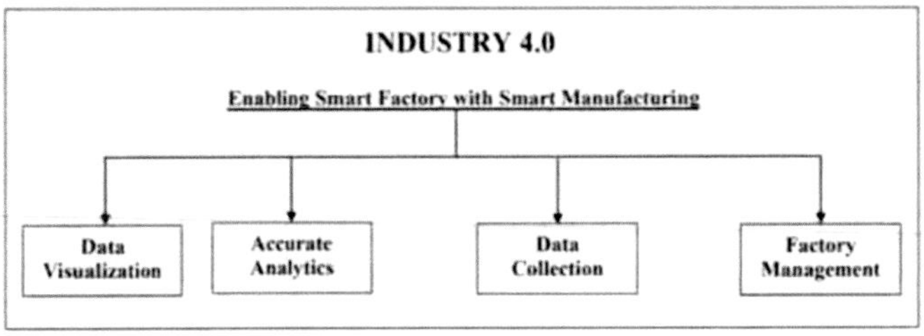

Fig. (1). Smart factory *vs* smart manufacturing.

History of the Industrial Revolution

The Industrial Revolution saw the shift from tiny, hand-operated cottage industries to new, mass-produced commodities built in factories powered by steam and water. Many of the technological advancements that contributed to the Industrial Revolution were developed in Britain, where it all started around the year 1760. In terms of employment, output value, and capital invested, the textile sector dominated the Industrial Revolution. The adoption of contemporary production techniques was also pioneered by the textile sector. The Industrial Revolution was a significant turning point in history that had some sort of impact on practically every element of daily life.

The rate of population growth experienced an unheard-of increase as a result of the Industrial Revolution. In addition, Great Britain rose to become the top commercial power in the globe, ruling a vast trading empire that included colonies in North America and the Caribbean and political sway over the Indian subcontinent. The Revolution is mainly divided into four phases, which are as follows [9 - 12].

Industrial Revolution-First

During the First Industrial Revolution, manual production methods were supplanted by machines using steam and water power. In Europe and the United States, it refers to the time between 1760 and 1820 or 1840 because it took a while for the new technology to be used.

The first industrial revolution had an impact on the textile sector, which was the first to adopt these innovations, as well as the iron industry, agriculture, and mining. It also had sociological repercussions, such as a growing middle class.

Industrial Revolution-Second

Massive railroad and telegraph networks were built during the 1871–1914 era, sometimes known as the Technological Revolution, to promote the faster exchange of people, ideas, and electricity. Companies were able to create modern production line thanks to increased electrification. During this period, significant economic growth and enhancement in productivity, which, together with the replacement of many factory workers by machines, led to an increase in unemployment [13, 14].

Industrial Revolution-Third

A slowdown in industrialization and technical advancement relative to preceding periods followed the end of the two world wars, which sparked the Third Industrial Revolution, often known as the Digital Revolution.

Industrial Revolution-Fourth

The Fourth Industrial Revolution is a trend in manufacturing technologies and processes that includes cloud computing, cognitive computing, artificial intelligence, industrial Internet of Things, cyber-physical systems, and IoT. The combination of machine learning and computer power enables machines to complete highly complex jobs, but they cannot replace deep expertise. Rather, they tend to be more efficient than people at executing repetitive activities. The phase of the Industrial Revolution is shown in Fig. (**2**).

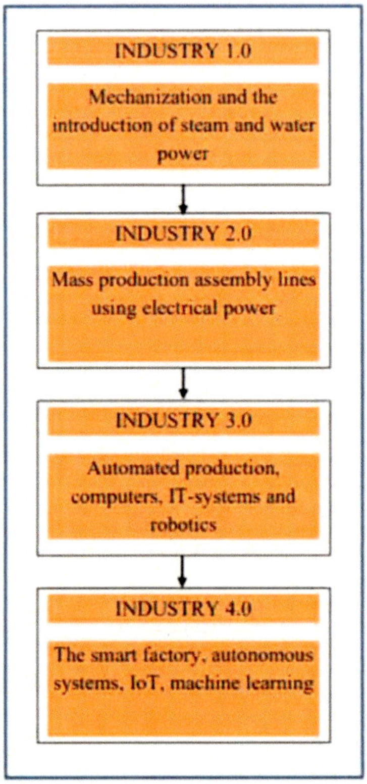

Fig. (2). Phases of the industrial revolution.

DESIGN PRINCIPLES OF INDUSTRY 4.0

Design concepts ought to be considered before putting Industry 4.0 into practice. The design concepts enable collaboration between Industry 4.0 components and comprehensive system adaption across the entire system. The following design principles were used in the application and implementation of Industry 4.0 [15, 16, 17].

- Agility: It refers to the system's adaptability to shifting requirements by upgrading or replacing discrete modules based on standardized software and hardware interfaces.

- Interoperability: It suggests that in order to construct a smart factory, the various parts of cyber-physical systems would communicate with one another *via* the Industrial Internet of Things and routine standardization procedures.
- Virtualization: Using simulation or augmented reality tools, virtualization enables system monitoring, new system adaptability, and system changes.
- Decentralization of decision: It is a critical phrase that describes how robots make decisions on their own and depend on learning from past experiences.
  Real-time data management: This involves the ability to gather and analyze data in real-time so that decisions can be made instantly and at any time.
- Service Orientation: Service orientation is the modification of the entire system to satisfy the needs of the client while taking into account both internal and external subsystems.
- Integrated business processes: By providing a communication and coordination mechanism aided by corporate data management services and connected networks, it serves as the interface between physical systems and software platforms.

TECHNOLOGIES IMPACTING PRODUCTION

Industry 4.0-related advanced manufacturing technologies are typically categorized into the following interrelated groups [18].

- Wireless technologies and sensors that allow the connection of people with machines and the subsequent collection of data and real-time decisions
- Network technologies that provide online capabilities
- Technology that processes data aimed at supporting the analysis of live data from production systems
- Technology that comprises equipment used in production.

THE PILLARS OF INDUSTRY 4.0

Data, artificial intelligence, technology, and communication are all combined in Industry 4.0 to create a productive industrial environment that is not just automated but also intelligent [19 - 23].

Internet of Things (IoT)

One essential element of smart manufacturing is the Internet of Things (IoT). Machines on the production line contain sensors with IP addresses that let them connect to other web-enabled equipment. This mechanization and connection enable extensive valuable data gathering, analysis, and exchange [24].

Cloud Computing

Cloud computing is a vital part of Industry 4.0 strategy. The full implementation of smart manufacturing requires the connectivity and integration of engineering, supply chain, production, sales and distribution, and services [25]. The cloud makes this possible. Additionally, the cloud can be used to process the normally vast amount of data being stored and evaluated more quickly and affordably. Cloud computing can dramatically reduce startup costs for small- and medium-sized manufacturers who can accurately analyze their demands and scale as their business grows [26].

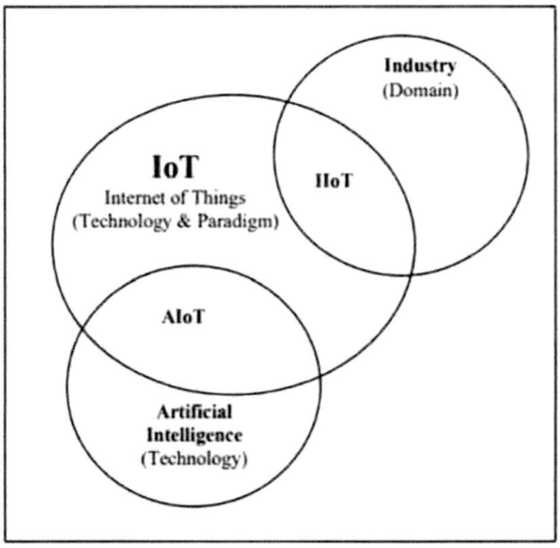

Fig. (3). AI, IoT and IIoT.

AI and Machine Learning

Thanks to AI and machine learning, manufacturing companies can fully capitalize on the wealth of information generated across all of their business units, on the factory floor, from partners, and from external sources. With the use of AI and machine learning, operations and business processes can become more visible, predictable, and automated. For instance, industrial equipment frequently malfunctions when a product is being produced. Utilizing the information acquired from these assets, businesses can perform predictive maintenance based on machine learning, enhancing uptime and effectiveness. Fig. (**3**) shows the relationship between IoT, IIoT and AI.

Edge Computing

Some data analysis must be done at the "edge" or where the data is generated, due to the requirements of real-time production operations. This shortens the time between the generation of data and the requirement for a response. For instance, the device may need to be used to take action in close to real time when a safety or quality concern is discovered. It can take too long to get data from the factory floor to the enterprise cloud and back, depending on how dependable the network is. Data stays close to its source when edge computing is used, lowering security threats.

Cyber Security

Cybersecurity and cyber-physical systems have not traditionally received significant attention from manufacturing companies [27]. The same operational equipment connectivity that improves industrial processes also creates new entry points for malicious software and unwelcomed intrusions [28]. When experiencing a digital transformation to Industry 4.0, it is necessary to take into account a cybersecurity strategy that involves both IT and OT equipment.

Digital Twin

A digital twin makes use of real-world data to create a simulation *via* a computer program that predicts how a product or process will perform. These programs are easily integrated into the Internet of Things (IoT) and artificial intelligence (AI), as well as software analytics, all in a bid to enhance output.

Cyber-Physical Systems

The cyber-physical systems are generally understood to be systems in which hardware and software components are seamlessly linked to carry out predetermined tasks. Fig. (**4**) shows the relationship between Industry 4.0 and different emerging technologies.

DIGITAL TWIN APPLICATIONS FOR MANUFACTURING

A precise virtual replica of an item, product, piece of machinery, person, procedure, supply chain, or even the entire company ecosystem is called a digital twin. It is produced using information obtained from sensor technologies connected to or incorporated in the original object that is part of the Internet of Things (IoT). Engineers may monitor systems and simulate system dynamics using this cloud-connected data, which provides both structural and operational views of what occurs to the item in real time.

Fig. (4). Emerging technologies.

There are several ways that digital twins are used to improve manufacturing operations [29, 30 - 32]:

- Designing of Product
  Before spending money on a sturdy prototype, digital twins can be used as virtual prototypes during the design phase to evaluate various simulations or ideas. Lowering the number of iterations necessary to put the product into production saves time and money.
- Optimization Process
  A digital twin of the process can be built using sensors on a production line, and key performance indicators can be examined. Changes to the digital twin can aid with root-cause analysis, find new opportunities for production optimization, and reduce variations.
- Management of Quality
  To maintain the highest level of quality and prevent rework, monitoring and reacting to data from IoT sensors during production are crucial. Every step of the production process can be modeled by the digital twin to find out where variations arise or whether better materials or procedures can be applied.

- Supply Chain
  Digital twins are used by supply chains and logistics/distribution companies to measure and examine key performance metrics like packaging performance, fleet management, and route efficiency. They are particularly helpful for planning just-in-time or just-in-sequence production and examining distribution paths [33, 34].
- Collaboration of cross-discipline
  Digital twins make it simple to communicate operational data across disciplines, fostering collaboration, better communication, and quicker decision-making. Using the same data, engineering, manufacturing, sales, and marketing can all collaborate to make better decisions.

Cyber-Physical Systems

Cyber-Physical Systems are one of the several enabling elements of Industry 4.0's cyber-physical component. The spectrum of cyber-physical systems is extremely distinct, as implied by the name cyber-physical. Therefore, these are the systems that strongly integrate the physical and digital worlds [35]. Physical reality plays a significant role [36]. As a result, these systems incorporate interaction between the physical and digital worlds [37].

Systems that are embedded are cyber-physical systems.We can conceptually think of the cyber-physical system as the combination of the embedded system and the physical system [38].

These embedded systems are particularly well-liked today since they are enabling technologies tomake systems smarter. Therefore, embedded systems have specific computing, communication, and control capabilities, regardless of the environment in which they are employed. Hence, examples of places where there is contact with the physical world through various sensors and actuators are medical devices, many of which have embedded systems in them, transportation vehicles, and military systems [39].

Embedded systems have computational and processing capabilities. The cyber-physical system is built with the help of these systems. Thus, the physical systems and the physical space are present in the cyber-physical system.This is essentially the internet. The interaction between them turns this physical space into a cyber-physical system [40].

When we talk about physical space, we frequently refer to various real objects, like people, homes, factories, cars, buildings, and so forth. Additionally, there are other items as well, such as many other household items or even items used in specific sectors, such as cell phones, cameras, washing machines, and refrigerators. Each of these unique items has a particular function in the physical world. These things essentially sense information [41]. They use specific communication devices after sensing various information in the environment where they are functioning. These data are transmitted further. As a result, after the sensing procedure is complete, the detected data is delivered for additional processing. Thus, this processing takes place across several servers, server companies, and clouds. This processing takes place in cyberspace, and the results are then transported back to the actual world to activate various gadgets based on their findings [42].

Cyber-Physical System (CPS) is a mechanism that is closely integrated with the internet and its users and is managed or controlled by computer-based algorithms. It is an engineering system that is based and depends on the seamless fusion of physical elements and computational algorithms. Cyber refers to discrete and logical calculation, communication, and control. Physical refers to organic and artificial systems that follow the principles of physics and function continuously. Cyber-Physical Systems are computing and communication systems that connect with the physical world [43, 44].

In the manufacturing sector, a CPS is a physical object that can connect to the internet, like a compressor or pump, and is incorporated with the computer control system's elements like sensors and actuators. Such an IP-assigned entity is capable of self-monitoring, information generation on its own operation, and communication with the outer world or perhaps with other linked entities. It is a self-governing, independent system. As the Fourth Industrial Revolution develops over the coming years, industrial firms will increasingly rely on the principles of CPS to help them gain operational excellence by enhancing productivity, effectively using all resources, satisfying customers, and increasing shareholder value [45 - 47]. In the third phase of the Industrial Revolution, which was marked by the automation of industrial production processes through the widespread usage of electronics and computers, the groundwork for Industry 4.0 was laid. Electronics and computers have become more widely used, which has led to price reductions, computer downsizing and ruggedness, and a significant rise in computing and processing capacity. The convergence of digital and communication technologies started and we saw the beginning of the Industrial Internet of Things (IIoT), edge/cloud computing, artificial intelligence (AI), autonomous robotics, and other technologies [48]. Manufacturing will change as a result of the combination of digital and industrial technologies, which will also contribute to the realization of the goal of Industry 4.0 [49].

The fundamental tenet of Industry 4.0 is that a manufacturing company can increase efficiency, productivity, and the autonomous operation of production processes by ensuring that machines/plant equipment, logistics systems, work-i--progress components, and such other elements directly communicate with one another to achieve collaboration. If a manufacturing company wants to connect its vision with Industry 4.0, it must take steps to convert existing physical entities into cyber-physical systems [50]. Fig. (**5**) shows the basic block diagram of the Cyber-Physical System.

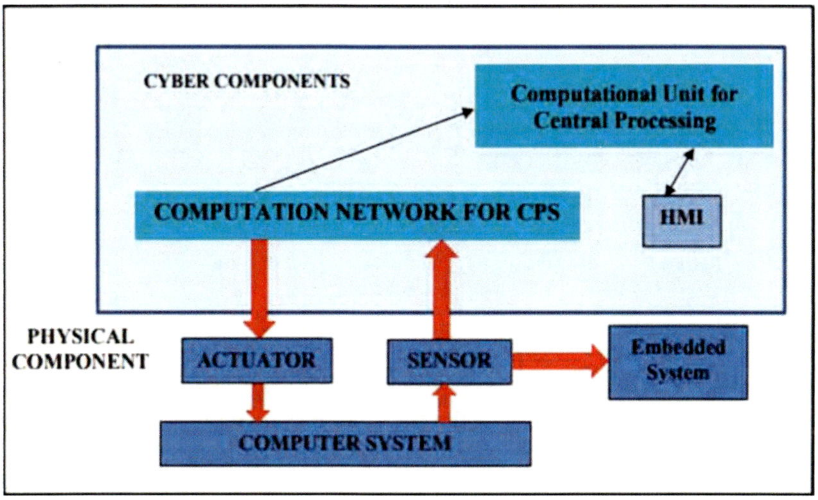

Fig. (5). Block diagram of the cyber-physical system.

THE BASIC CHARACTERISTICS OF INDUSTRY 4.0 [51]

- Switching to robototronics from manual labor, which ensures that all production operations are automated.
- Modernization of transportation and logistics infrastructure brought on by the widespread usage of unmanned vehicles.
- A rise in the complexity and accuracy of manufactured technological goods, as well as the manufacturing of new building materials, as a result of better production technologies.
- Due to advancements in production technology, the complexity and accuracy of manufactured technical items have increased, and new construction materials are being produced.
- The Internet of Things technology is being used to advance inter-machine communications and the self-management of physical systems.

FEATURES OF A SMART FACTORY

A manufacturing facility that integrates digital technology, smart computing, artificial intelligence, and big data with actual production processes and operations is referred to as a "smart factory". It promises possible effects on manufacturing and production, such as boosted productivity, decreased costs, decreased downtime, and less waste [52]. Why do we even need smart factories? Today's market is extremely competitive, which makes it difficult for our technology to advance. The adoption of these technologies will almost certainly result in increased production, efficiency, better monitoring of this equipment, and decreased downtime, processes, and systems overall. This will contribute to raising market competition. It should be clear to everyone that in order to compete, we need smart factories that can produce a lot of stuff in a short amount of time [53]. The last factor that is crucial is lowering the possibility of failure, which will have an impact on production. Consequently, lowering the risk of failure can be accomplished by using the various features of the smart factory [54]. Some major characteristics of the smart factory are as follows:

Networks and Technologies

Both traditional and smart factories use IT and electronic equipment, but smart factories integrate IoT devices with networks to connect and allow data exchanges that enable deeper data insights. Additionally, intelligent machines, robots, tools, and workpieces regulate and monitor the production processes in a smart factory, resulting in automation and self-optimization. As an illustration, embedded sensors provide sensor data analytics that aids in preventive maintenance, where a machine's failure warning is followed by an automatic repair signal.

Decision Making Process

Companies or manufacturers receive a variety of information based on previous data and patterns during the production process thanks to data analytics done by smart factories. Additionally, depending on data from human resources, sales, warehouses, and staff, they can improve their production process by making better decisions [55, 56].

Supply Chain

Supply chains are made more effective and transparent by smart factories. Sharing production data with the suppliers is a part of it. This enables businesses or manufacturers to make better predictions or to have faster shipment or delivery times. The potential of digital transformation as a remedy for numerous enterprises and industries has been realized. It follows that smart factories will eventually become the “factories of the future”, as was formerly thought [57, 58].

Production

Mass manufacturing is the primary goal of traditional factories, while mass customization is the focus of smart factories. Because smart factories use a more flexible production and logical system, the products they make are more innovative and one-of-a-kind and have shorter product life cycles. In addition, a smart factory's more effective resource use makes it possible for it to produce in a sustainable way over a longer period of time than a typical factory [59].

Now, the primary benefits of operating a smart factory include lowering overall production costs, raising productivity and product efficiency, enhancing the effectiveness of the surfaces being used, and enhancing the effectiveness of the manufacturing processes. So, incorporating or switching to smart factories can help attain total efficiency in all distinct parameters [60].

These are the various elements of a smart factory: smart machines, smart engineering, smart devices, and a smart manufacturing process. The other three elements are improving existing innovations by inventing new products and manufacturing techniques, and all of these things are possible with the aid of information technology. The factory as a whole can become smarter with the use of information technology [61]. This makes IT a crucial component, as its application in manufacturing can assist in making businesses, factories, and data analytics smarter. When you integrate IoT solutions, sensors, and actuators into these smart factories, a ton of data is generated that needs to be evaluated in order to make the factories smarter.

CHALLENGES AND OPPORTUNITIES IN THE IMPLEMENTATION OF INDUSTRY 4.0

The challenges the industry faces while implementing Industry 4.0 have different aspects. Task automation will expand as a result of Industry 4.0, thus workers must be ready to take on new responsibilities. The same is true of engineering education, which has a great deal of promise for developing future professionals and educating them about emerging technological trends and opportunities. The emergence of new and diverse technology in the 1980s changed the mechanical industries into highly automated sectors. These sectors respond and adapt quickly to changing consumer demands [62, 62].

The technologies of Industry 4.0 have given manufacturers fantastic chances to learn important lessons, reorganize workflows, and establish more effective and adaptable operations. However, the advanced manufacturing industry's reliance on connectivity and intelligence exposes businesses to dangers from numerous angles [63, 64].

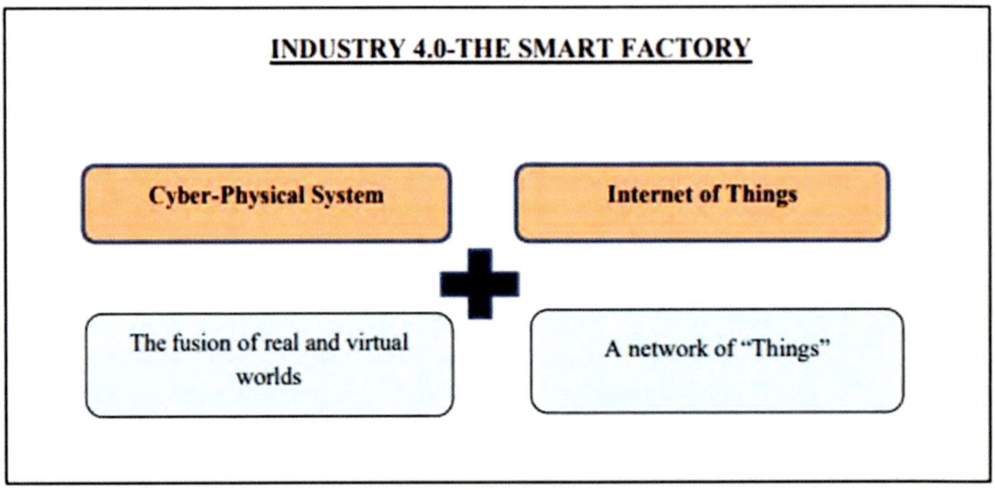

Fig. (6). The interaction between real and virtual words.

In the implementation of Industry 4.0 as a novel concept, the industries may face some challenges, such as low availability of standard and benchmarked processes, incompatibility of the new technology with the workforce, and chances of error in data processing and upgradation of the existing apparatus [65, 66].

CONCLUSION AND FUTURE SCOPE

Industry 4.0 not only improves the productivity and quality of the production process, but it also makes upgrades and preventive maintenance possible, reducing downtime and capital expenditure over time. Manufacturing firms' efficiency and competitiveness can be improved with the aid of Industry 4.0. However, there is a major challenge of deployment. Firms will find it simpler to successfully deploy Industry 4.0 if they can convince their staff members of the advantages of digital technologies and give them confidence in their ability to apply them. Another important point is the high cost of implementation, maintenance, and training. These are the significant obstacles to the adoption of Industry 4.0. If these parameters are controlled, then naturally, industry 4.0 will be beneficial from different aspects for industries.

Industry 4.0 is anticipated to bring about improvements in the production process as well as certain changes to the social, economic, environmental, and political systems. The future of work and job concerns undoubtedly rank at the top of these trends.

REFERENCES

[1] V.S. Narwane, R.D. Raut, V.S. Yadav, and A.R. Singh, "Barriers in sustainable industry 4.0: A case study of the footwear industry", *Int. J. Sustain. Eng.*, vol. 14, no. 3, pp. 175-189, 2021. [http://dx.doi.org/10.1080/19397038.2020.1836065]

[2] J.R. Reagan, and M. Singh, "Management 4.0 : Cases and Methods for the 4th Industrial Revolution", [http://dx.doi.org/10.1007/978-981-15-6751-3]

[3] S.K. Das, S. Samal, P. Ranjan, and S.P. Das, "An Efficient Data Mining Technique for an Intrusion Detection System in Network", In: *Constraint Decision-Making Systems in Engineering.* IGI Global, 2023, pp. 1-17. [http://dx.doi.org/10.4018/978-1-6684-7343-6.ch001]

[4] S. Majumder, and N. Dey, *AI-empowered knowledge management.* Springer, 2022. [http://dx.doi.org/10.1007/978-981-19-0316-8]

[5] R. Rai, M.K. Tiwari, D. Ivanov, and A. Dolgui, "Machine learning in manufacturing and industry 4.0 applications", *Int. J. Prod. Res.*, vol. 59, no. 16, pp. 4773-4778, 2021. [http://dx.doi.org/10.1080/00207543.2021.1956675]

[6] A. Sarker, M. Mozaffari Kermani, and R. Azarderakhsh, "Efficient Error Detection Architectures for Postquantum Signature Falcon's Sampler and KEM SABER", *IEEE Trans. Very Large Scale Integr. (VLSI) Syst.*, vol. 30, no. 6, pp. 794-802, 2022. [http://dx.doi.org/10.1109/TVLSI.2022.3156479]

[7] S. V Anil Kumar G. Bawge, and V. Kumar, "An Overview of Industrial Revolution and Technology of Industrial 4.0", *(IJRES).*, vol. 9, ISSN, 2021.

[8] M. Mozaffari-Kermani, *Reliable and high-performance hardware architectures for the Advanced Encryption Standard/Galois Counter Mode.* The University of Western Ontario: Canada, 2011.

[9] G-J. Cheng, L-T. Liu, X-J. Qiang, and Y. Liu, "Industry 4.0 development and application of intelligent manufacturing", *International conference on information system and artificial intelligence (ISAI),*, pp. 407-410, 2016. [http://dx.doi.org/10.1109/ISAI.2016.0092]

[10] E. Oztemel, and S. Gursev, "Literature review of Industry 4.0 and related technologies", *J. Intell. Manuf.*, vol. 31, no. 1, pp. 127-182, 2020. [http://dx.doi.org/10.1007/s10845-018-1433-8]

[11] G. Culot, G. Nassimbeni, G. Orzes, and M. Sartor, "Behind the definition of Industry 4.0: Analysis and open questions", *Int. J. Prod. Econ.*, vol. 226, p. 107617, 2020. [http://dx.doi.org/10.1016/j.ijpe.2020.107617]

[12] M. A. K. Bahrin, M. F. Othman, N. H. N. Azli, and M. F. Talib, "Industry 4.0: A review on industrial automation and robotic", *J Teknol,* vol. 78, p. 6–13, 2016.

[13] A. Sarker, A.C. Canto, M.M. Kermani, and R. Azarderakhsh, "Error detection architectures for hardware/software co-design approaches of number-theoretic transform", *IEEE Trans. Comput. Aided Des. Integrated Circ. Syst.,* 2022.

[14] M.M. Kermani, and R. Azarderakhsh, "Integrating emerging cryptographic engineering research and security education", *2015 ASEE Annual Conference & Exposition,* pp. 26-989, 2015.

[15] A. Ustundag, and E. Cevikcan, *Industry 4.0: managing the digital transformation.* Springer, 2017.

[16] L. Kanski, and J. Pizon, "The impact of selected components of industry 4.0 on project management", *J. Innov. Knowl.,* vol. 8, no. 1, p. 100336, 2023. [http://dx.doi.org/10.1016/j.jik.2023.100336]

[17] C. Salkin, M. Oner, A. Ustundag, and E. Cevikcan, "A conceptual framework for Industry 4.0", *Industry 4.0: managing the digital transformation,* pp. 3-23, 2018. [http://dx.doi.org/10.1007/978-3-319-57870-5_1]

[18] S.K. Das, N. Dey, R. González Crespo, and E. Herrera-Viedma, "A non-linear multi-objective technique for hybrid peer-to-peer communication", *Inf. Sci.,* vol. 629, pp. 413-439, 2023. [http://dx.doi.org/10.1016/j.ins.2023.01.117]

[19] A.A. Wagire, A.P.S. Rathore, and R. Jain, "Exploration of Pillars of Industry 4.0 Using Latent Semantic Analysis Technique", In: *Smart Innovation.* Systems and Technologies, 2020. [http://dx.doi.org/10.1007/978-981-15-1616-0_69]

[20] S. Luthra, A. Kumar, E.K. Zavadskas, S.K. Mangla, and J.A. Garza-Reyes, "Industry 4.0 as an enabler of sustainability diffusion in supply chain: an analysis of influential strength of drivers in an emerging economy", *Int. J. Prod. Res.,* vol. 58, no. 5, pp. 1505-1521, 2020. [http://dx.doi.org/10.1080/00207543.2019.1660828]

[21] S. Ramakrishna, A. Ngowi, H.D. Jager, and B.O. Awuzie, "Emerging industrial revolution: Symbiosis of Industry 4.0 and circular economy: The role of universities", *Sci. Technol. Soc.,* vol. 25, no. 3, pp. 505-525, 2020. [http://dx.doi.org/10.1177/0971721820912918]

[22] A.G. Frank, L.S. Dalenogare, and N.F. Ayala, "Industry 4.0 technologies: Implementation patterns in manufacturing companies", *Int. J. Prod. Econ.,* vol. 210, pp. 15-26, 2019. [http://dx.doi.org/10.1016/j.ijpe.2019.01.004]

[23] L.S. Dalenogare, G.B. Benitez, N.F. Ayala, and A.G. Frank, "The expected contribution of Industry 4.0 technologies for industrial performance", *Int. J. Prod. Econ.,* vol. 204, pp. 383-394, 2018. [http://dx.doi.org/10.1016/j.ijpe.2018.08.019]

[24] J. Kaur, A.C. Canto, M.M. Kermani, and R. Azarderakhsh, "A comprehensive survey on the implementations, attacks, and countermeasures of the current NIST lightweight cryptography standard", *2304.06222,* 2023.

[25] E. Dubrova, K. Ngo, J. Gärtner, and R. Wang, "Breaking a fifth-order masked implementation of crystals-kyber by copy-paste", *Proceedings of the 10th ACM Asia Public-Key Cryptography Workshop,* pp. 10-20, 2023. [http://dx.doi.org/10.1145/3591866.3593072]

[26] A.C. Canto, J. Kaur, M.M. Kermani, and R. Azarderakhsh, "Algorithmic security is insufficient: A comprehensive survey on implementation attacks haunting post-quantum security", *2305.13544,* 2023.

[27] M.M. Kermani, E. Savas, and S.J. Upadhyaya, "Guest editorial: Introduction to the special issue on emerging security trends for deeply-embedded computing systems", *IEEE Trans. Emerg. Top. Comput.,* vol. 4, no. 3, pp. 318-320, 2016. [http://dx.doi.org/10.1109/TETC.2015.2482778]

[28] A. Cintas-Canto, J. Kaur, M. Mozaffari-Kermani, and R. Azarderakhsh, "ChatGPT vs. Lightweight Security: First Work Implementing the NIST Cryptographic Standard ASCON", *2306.08178,* .

[29] A. Sharma, E. Kosasih, J. Zhang, A. Brintrup, and A. Calinescu, "Digital Twins: State of the art theory and practice, challenges, and open research questions", *J. Ind. Inf. Integr.,* vol. 30, p. 100383, 2022. [http://dx.doi.org/10.1016/j.jii.2022.100383]

[30] G. Dalmarco, F.R. Ramalho, A.C. Barros, and A.L. Soares, "Providing industry 4.0 technologies: The case of a production technology cluster", *J. High Technol. Manage. Res.,* vol. 30, no. 2, p. 100355, 2019. [http://dx.doi.org/10.1016/j.hitech.2019.100355]

[31] C. Bai, P. Dallasega, G. Orzes, and J. Sarkis, "Industry 4.0 technologies assessment: A sustainability perspective", *Int. J. Prod. Econ.,* vol. 229, p. 107776, 2020. [http://dx.doi.org/10.1016/j.ijpe.2020.107776]

[32] A. Raj, G. Dwivedi, A. Sharma, A.B. Lopes de Sousa Jabbour, and S. Rajak, "Barriers to the adoption of industry 4.0 technologies in the manufacturing sector: An inter-country comparative perspective", *Int. J. Prod. Econ.,* vol. 224, p. 107546, 2020.

[http://dx.doi.org/10.1016/j.ijpe.2019.107546]

[33] D. Ivanov, A. Dolgui, and B. Sokolov, "Cloud supply chain: Integrating Industry 4.0 and digital platforms in the "Supply Chain-as-a-Service"", *Transp. Res., Part E Logist. Trans. Rev.*, vol. 160, p. 102676, 2022.
[http://dx.doi.org/10.1016/j.tre.2022.102676]

[34] V.S. Yadav, A.R. Singh, R.D. Raut, S.K. Mangla, S. Luthra, and A. Kumar, "Exploring the application of Industry 4.0 technologies in the agricultural food supply chain: A systematic literature review", *Comput. Ind. Eng.*, vol. 169, p. 108304, 2022.
[http://dx.doi.org/10.1016/j.cie.2022.108304]

[35] M.M. Kermani, R. Azarderakhsh, and J. Xie, "Error detection reliable architectures of Camellia block cipher applicable to different variants of its substitution boxes", *2016 IEEE Asian Hardware-Oriented Security and Trust (AsianHOST).*, pp. 1-6, 2016.

[36] B. Koziel, A. Jalali, R. Azarderakhsh, D. Jao, and M. Mozaffari-Kermani, ""NEON-SIDH: Efficient implementation of supersingular isogeny Diffie-Hellman key exchange protocol on ARM," in Cryptology and Network Security: 15th International Conference, CANS 2016, Milan, Italy, November 14-16, 2016", *Proceedings,* vol. 15, pp. 88-103, 2016.

[37] M.E. Karim, and V.V. Phoha, "Cyber-physical systems security", *Applied Cyber-Physical Systems,* pp. 75-83, 2014.
[http://dx.doi.org/10.1007/978-1-4614-7336-7_7]

[38] M.A. Moktadir, S.M. Ali, S. Kusi-Sarpong, and M.A.A. Shaikh, "Assessing challenges for implementing Industry 4.0: Implications for process safety and environmental protection", *Process Saf. Environ. Prot.*, vol. 117, pp. 730-741, 2018.
[http://dx.doi.org/10.1016/j.psep.2018.04.020]

[39] Y. Ashibani, and Q.H. Mahmoud, "Cyber physical systems security: Analysis, challenges and solutions", *Comput. Secur.*, vol. 68, pp. 81-97, 2017.
[http://dx.doi.org/10.1016/j.cose.2017.04.005]

[40] J.P.A. Yaacoub, O. Salman, H.N. Noura, N. Kaaniche, A. Chehab, and M. Malli, "Cyber-physical systems security: Limitations, issues and future trends", *Microprocess. Microsyst.*, vol. 77, p. 103201, 2020.
[http://dx.doi.org/10.1016/j.micpro.2020.103201] [PMID: 32834204]

[41] S.K. Das, *Smart design and its applications. Challenges and techniques.* Nature-Inspired Computing for Smart Application Design, 2021, pp. 1-6.

[42] Y. Zacchia Lun, A. D'Innocenzo, F. Smarra, I. Malavolta, and M.D. Di Benedetto, "State of the art of cyber-physical systems security: An automatic control perspective", *J. Syst. Softw.*, vol. 149, pp. 174-216, 2019.
[http://dx.doi.org/10.1016/j.jss.2018.12.006]

[43] W. Lidong, and W. Guanghui, "Big data in cyber-physical systems, digital manufacturing and industry 4.0", *Int. J. Eng. Manufact.*, vol. 6, no. 4, pp. 1-8, 2016.
[http://dx.doi.org/10.5815/ijem.2016.04.01]

[44] B. Bagheri, S. Yang, H.A. Kao, and J. Lee, "Cyber-physical systems architecture for self-aware machines in industry 4.0 environment", *IFAC-PapersOnLine,* vol. 48, no. 3, pp. 1622-1627, 2015.
[http://dx.doi.org/10.1016/j.ifacol.2015.06.318]

[45] A.A.F. Saldivar, Y. Li, W. Chen, Z. Zhan, J. Zhang, and L.Y. Chen, "Industry 4.0 with cyber-physical integration: A design and manufacture perspective", *2015 21st international conference on automation and computing (ICAC),*, pp. 1-6, 2015.
[http://dx.doi.org/10.1109/IConAC.2015.7313954]

[46] N. Jazdi, "Cyber physical systems in the context of Industry 4.0", *2014 IEEE international conference on automation, quality and testing, robotics.*, pp. 1-4, 2014.

[47] Y. Lu, "Cyber physical system (CPS)-based industry 4.0: A survey", *J. Indu. Integ. Manag.*, vol. 2, no. 3, p. 1750014, 2017. [http://dx.doi.org/10.1142/S2424862217500142]

[48] A.A. Serag, and M.M. Daoud, "A proposed Framework for Studying the Impact of Cybersecurity on Accounting Information to Increase Trust in The Financial Reports in the Context of Industry 4.0: An Event", *Impact and Response Approach.*, vol. 42, no. 1, pp. 20-61, 2022.

[49] J. Lee, B. Bagheri, and H.A. Kao, "A Cyber-Physical Systems architecture for Industry 4.0-based manufacturing systems", *Manuf. Lett.*, vol. 3, pp. 18-23, 2015. [http://dx.doi.org/10.1016/j.mfglet.2014.12.001]

[50] P.J. Mosterman, and J. Zander, "Industry 4.0 as a cyber-physical system study", *Soft. Syst. Model.*, vol. 15, no. 1, pp. 17-29, 2016. [http://dx.doi.org/10.1007/s10270-015-0493-x]

[51] S. Heng, "Industry 4.0: Upgrading of Germany's industrial capabilities on the horizon", *SSRN,* p. 2656608, 2014.

[52] C. Nagpal, P.K. Upadhyay, S.S. Hussain, A.C. Bimal, and S. Jain, "IIoT based smart factory 4.0 over the cloud", *2019 International Conference on Computational Intelligence and Knowledge Economy (ICCIKE),* pp. 668-673, 2019. [http://dx.doi.org/10.1109/ICCIKE47802.2019.9004413]

[53] N. Dey, *Data-Driven Approach for Bio-medical and Healthcare.* Springer Nature, 2022.

[54] A.K. Dass, and S.K. Das, "An Optimal Data Communication System in Wireless Sensor Network Using Artificial Neural Network", In: *Constraint Decision-Making Systems in Engineering.* IGI Global, 2023, pp. 120-134. [http://dx.doi.org/10.4018/978-1-6684-7343-6.ch007]

[55] C.L. Garay-Rondero, J.L. Martinez-Flores, N.R. Smith, S.O.C. Morales, and A. Aldrette-Malacara, "Digital supply chain model in Industry 4.0", *J. Manuf. Tech. Manag.*, 2020.

[56] H. Fatorachian, and H. Kazemi, "Impact of Industry 4.0 on supply chain performance", *Prod. Plann. Contr.*, vol. 32, no. 1, pp. 63-81, 2021. [http://dx.doi.org/10.1080/09537287.2020.1712487]

[57] B. Tjahjono, C. Esplugues, E. Ares, and G. Pelaez, "What does industry 4.0 mean to supply chain?", *Procedia Manuf.*, vol. 13, pp. 1175-1182, 2017. [http://dx.doi.org/10.1016/j.promfg.2017.09.191]

[58] S. Bag, A. Telukdarie, J.H.C. Pretorius, and S. Gupta, "Industry 4.0 and supply chain sustainability: framework and future research directions", *Benchmarking,* p. BIJ-03-2018-0056, 2018. [http://dx.doi.org/10.1108/BIJ-03-2018-0056]

[59] S. Mukhuty, A. Upadhyay, and H. Rothwell, "Strategic sustainable development of Industry 4.0 through the lens of social responsibility: The role of human resource practices", *Bus. Strategy Environ.*, vol. 31, no. 5, pp. 2068-2081, 2022. [http://dx.doi.org/10.1002/bse.3008]

[60] N. Malik, S.N. Tripathi, A.K. Kar, and S. Gupta, "Impact of artificial intelligence on employees working in industry 4.0 led organizations", *Int. J. Manpow.*, vol. 43, no. 2, pp. 334-354, 2022. [http://dx.doi.org/10.1108/IJM-03-2021-0173]

[61] Y. Srivastava, S. Ganguli, S. Suman Rajest, and R. Regin, "Smart HR Competencies and Their Applications in Industry 4.0", *A Fusion of Artificial Intelligence and Internet of Things for Emerging Cyber Systems,* pp. 293-315, 2022. [http://dx.doi.org/10.1007/978-3-030-76653-5_16]

[62] J. Tupa, J. Simota, and F. Steiner, "Aspects of risk management implementation for Industry 4.0", *Procedia Manuf.*, vol. 11, pp. 1223-1230, 2017.

[http://dx.doi.org/10.1016/j.promfg.2017.07.248]

[63] S.S. Kamble, A. Gunasekaran, and R. Sharma, "Analysis of the driving and dependence power of barriers to adopt industry 4.0 in Indian manufacturing industry", *Comput. Ind.,* vol. 101, pp. 107-119, 2018.
[http://dx.doi.org/10.1016/j.compind.2018.06.004]

[64] S. Luthra, and S.K. Mangla, "Evaluating challenges to Industry 4.0 initiatives for supply chain sustainability in emerging economies", *Process Saf. Environ. Prot.,* vol. 117, pp. 168-179, 2018.
[http://dx.doi.org/10.1016/j.psep.2018.04.018]

[65] M. Lezzi, M. Lazoi, and A. Corallo, "Cybersecurity for Industry 4.0 in the current literature: A reference framework", *Comput. Ind.,* vol. 103, pp. 97-110, 2018.
[http://dx.doi.org/10.1016/j.compind.2018.09.004]

[66] D. Sanghavi, S. Parikh, and S.A. Raj, *Industry 4.0: Tools and implementation.* Management and Production Engineering Review, 2019.

CHAPTER 6

Emerging Technologies in Fintech: A Case Study

Keyurkumar Patel[1,*], Pujita Sunnapu[2] and Sunil Gautam[3]

[1] *School of IT, Artificial Intelligence and Cyber Security, Rashtriya Raksha University, Gandhinagar, Gujarat-382305, India*

[2] *Unitedworld School of Computational Intelligence, Karnavati University, Uvarsad, Gandhinagar – 382422, India*

[3] *Department of Computer Science and Engineering, Institute of Technology, Nirma University, Ahmedabad, Gujarat-382481, India*

Abstract: The Financial Technology (FinTech) industry has been playing a pivotal role in driving modern day's economics, social aspects, technology, and many more areas. FinTech is majorly inspired, motivated, and empowered by Data Science and Artificial Intelligence Methodologies (DSAIM). With emerging technology, the smart FinTech industry has revolutionised economic and financial businesses, service industries, and systems. The global research communities have made significant progress in smart FinTech for Banking Tech, Trade Tech, InsurTech, Wealth Tech, Pay Tech, Risk Tech, Cryptocurrencies, Digital Payment Systems, and Blockchain using DSAIM. In this review paper, we narrow down the overview of smart financial businesses, their complex challenges, and the entire smart FinTech ecosystem. The DSAIM enables smart FinTech and poses some research problems among global academic and researcher communities.

Keywords: AI, AI in finance, AI in fintech, Blockchain, DSAIM, Fintech, Financial service.

INTRODUCTION

The financial sector has witnessed significant delivery changes as a result of automation in the span of the last 10 years. This transformation is distinguished by increased connectivity and faster information processing on the part of both customers and financial organisations such as banking institutions (back-office), insurance providers, brokerages, and many other firms involved in financial services. In the 21st century, finance has become an increasingly popular interactive stage for data science and artificial intelligence methodologies

* **Corresponding author Keyurkumar Patel:** School of IT, Artificial Intelligence and Cyber Security, Rashtriya Raksha University, Gandhinagar, Gujarat-382305, India; E-mail: pkeyur.436@gmail.com

Santosh Kumar Das, Soumi Majumder and Nilanjan Dey (Eds.)

(DSAIM). DSAIM advances the development of previously unknown and far more sophisticated, effective, convenient, personalised, comprehensible, secure, and assertive financial products and services in addition to the efficiency, expenditure, consumer experience, mitigating risk, regulatory framework, and protection of current financial systems. Here, DSAIM widely refers to (1) Traditional AI areas including expert systems, recommended systems, pattern recognition, and human emotion recognition (2) More advanced DSAIM areas such as intelligent transportation, intelligent machine, advanced data analytics and learning, discovery of knowledge, computational intelligence, behaviour analysis, and social network [1]. In addition, foundational fields like mathematical and statistical modelling play a crucial and necessary role in allowing FinTech. FinTech refers to the financial industry pioneers and disruptors who make use of prevalent technologies such as the Internet and automated data processing. These new solutions can come from startups, defined providers of financial services, or even established technology companies.

Mobile phones, which connect gadgets functioning from home, public spaces, offices, colleges, and schools, are rampant in the current generation. Companies in the banking, finance and insurance sectors can quickly adopt IoT technology. The financial services and banking industries have developed a new method for gathering crucial client data utilising IoT sensor devices and mobile phones. As compared to current financial technology, fintech is similar to electric car innovation in that it deals with finding out the best trading investment, and asset valuation and offers new and better accounting methods. A significant strategy in financial services is the Internet of Things. The internet-connected IoT transmits and receives information that is saved in the cloud when it is linked to the concerned bank. The bank has given these devices authorization to share data on the cloud, which enables all customers to examine account information and provide access while using mobile devices. Messages and awaiting work alerts are the simplest way to communicate with the consumer and also to provide personal information [2].

Big data is the term used to describe the enormous quantities of unstructured and structured information that is produced by people, companies, and other organisations. Big data is essential in the field of fintech because it enables businesses to examine massive volumes of data and acquire an understanding of consumer behaviour, market dynamics, and risk management. Big data is used by fintech organisations to handle and analyse massive volumes of financial data in real time, enabling them to make wise decisions fast and effectively. These businesses can use big data to find trends and patterns that aid in forecasting market trends, dangers, and opportunities. Fintech companies may improve client experiences, expand their product offerings, and improve service by utilising big data analytics [3].

The adoption of data science in financial services has been substantial. The main aim of this paper is to review existing literary works regarding the application of DSAI for multiple digital finance subject areas and to offer an overview of the various existing use cases and techniques of DSAI for future practitioners and researchers to use.

Motivation

The motivation behind blockchain technology is to provide a secure and transparent way of conducting transactions and storing data without the need for intermediaries. The importance of blockchain lies in its ability to increase efficiency and reduce costs, as well as to improve security and privacy.

The motivation behind AI in finance is to automate mundane tasks, reduce operational costs and provide more personalized experiences to customers. The importance of AI lies in its ability to analyze vast amounts of data and provide insights to inform decision-makers, as well as in its ability to improve customer experience.

The motivation behind digital payments is to provide an easy, fast and convenient way for people to make transactions without the need for cash or checks. The importance of digital payments lies in its ability to increase financial inclusion and improve access to financial services for everyone, especially those who are unbanked or underbanked.

The motivation behind robo-advisors is to provide low-cost investment advice and portfolio management to investors, reducing the need for human financial advisors. The importance of robo-advisors lies in their ability to democratize access to investment advice and to provide more efficient and personalized investment solutions to customers.

Advantages:

Blockchain: Some advantages of blockchain technology include increased transparency, reduced costs, improved security and privacy, and increased efficiency.

Artificial Intelligence: Some advantages of AI in finance include increased efficiency, improved customer experience, and better risk management.

Digital Payments: Some advantages of digital payments include increased financial inclusion, improved access to financial services, and reduced fraud and corruption.

Robo-Advisors: Some advantages of robo-advisors include lower costs, increased accessibility, and more personalized investment advice.

Limitations:

Blockchain: Some limitations of blockchain technology include scalability issues, regulatory challenges, and a lack of understanding about the technology among potential users.

Artificial Intelligence: Some limitations of AI in finance include ethical concerns such as bias and discrimination, a lack of transparency in decision-making, and potential job losses.

Digital Payments: Some limitations of digital payments include limited infrastructure and access in some regions, security and privacy concerns, and resistance to change from some individuals and businesses.

Robo-Advisors: Some limitations of robo-advisors include a lack of human interaction, the possibility of algorithmic bias, and the need for a certain level of financial literacy among users.

LITERATURE REVIEW

Fintech, or financial technology, has developed quickly over the past few years and is still expanding quickly. Blockchain, artificial intelligence, machine learning, big data, and cloud computing are a few of the newer finance technologies. The way financial services are delivered is changing as a result of these technologies, which are also having a big impact on the traditional financial sector.

Blockchain technology has the ability to completely change how money is transacted. Without the use of middlemen, it enables secure, open, and decentralised transactions. Among other applications, the technology can be used to create digital wallets, decentralised exchanges, and currencies [4].

In the financial industry, artificial intelligence (AI) and machine learning (ML) are frequently used for a range of activities, such as credit scoring, fraud detection, and investment decision-making. Financial institutions may use AI and ML algorithms to process vast volumes of data fast and accurately, allowing them to make better decisions and provide better customer service [5].

Big data is becoming more significant in fintech. Big data analytics is being used by financial institutions to understand consumer behaviour, market trends, and risk management. The development of more individualised financial services and products, as well as a better understanding of consumer needs and preferences, are both possible with the use of this data.

The delivery of financial services is also altering as a result of cloud computing. Financial organisations can lower costs, increase scalability, and improve security by shifting their infrastructure and applications to the cloud. Due to this, many fintech businesses have been able to create novel products and services that would not have been feasible in the conventional banking industry.

The results of the literature research demonstrate that the financial industry is being significantly impacted by emerging technology. Financial institutions are now able to enhance their services and better satisfy the needs of their clients thanks to the implementation of these technologies. These technologies do, however, also present difficulties that must be resolved, such as security and privacy issues.

FINTECH ECOSYSTEM

We must first examine the ecosystem in order to comprehend the dynamics of coordination and competition in fintech innovation. The success of the fintech sector needs a robust symbiotic fintech ecosystem. The participants in a fintech ecosystem include business owners, the government, and financial institutions [6].

- Fintech startups (such as payment, wealth management, lending, crowdfunding, capital market, and insurance fintech companies).
- Technological development (such as big data analytics, cloud computing, IoT, blockchain, cryptocurrency, and social media developers).
- Government organisations (such as financial regulators and legislation).
- Financial service customers (such as individuals and institutions).
- Different financial institutions (such as traditional banks, insurance companies, and stock brokerage) are all categories of the financial sector.

These factors work together to foster innovation, boost the economy, encourage cooperation and competitiveness in the financial sector, and ultimately benefit financial industry consumers. The five components of the fintech ecosystem are depicted in Fig. (**1**).

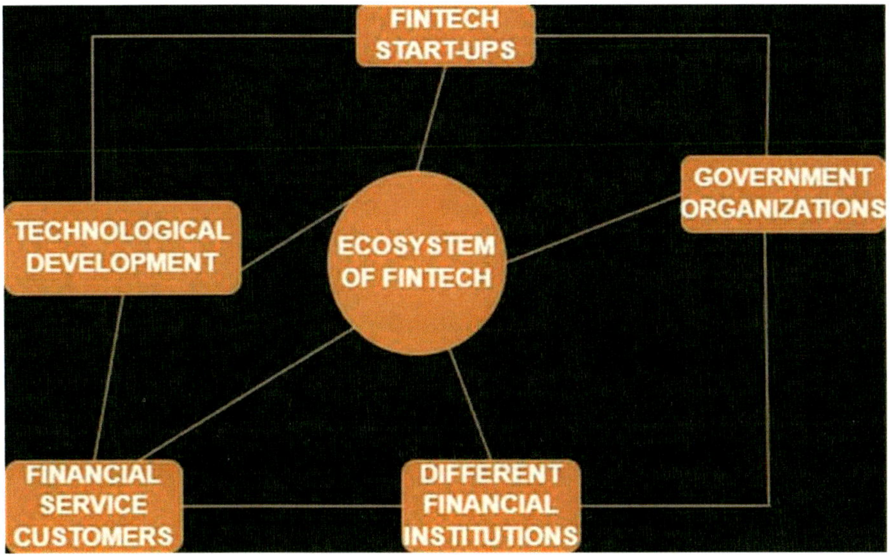

Fig. (1). - FinTech Ecosystem.

Fintech startups are at the epicentre of the ecosystem. By having to incur lower operating costs, focusing on more specialised markets, and offering more individualised services than conventional financial companies, these companies, which are primarily entrepreneurial, have driven significant innovations in the fields of payment, wealth management, lending, crowdfunding, capital markets, and insurance.

They are responsible for the Internet financial services trend, which has caused major disruptions for traditional financial institutions. Since traditional banks are at a disadvantage in this position, the capacity to unblock services is one of the key factors influencing the development of the fintech sector [7].

FINTECH BUSINESS CHALLENGES

Fintech firms are on a path to broaden and establish a reputation in the industry. Financial technology is evolving and automating business processes and services. However, these challenges grow our imaginations in new directions and foster unprecedented growth.

FinTech Business Areas:

Based on DSAIM techniques, these areas cover almost every aspect of a FinTech business in general. The mentioned major business areas in FinTech are highlighted here:

- Banking
- Investing
- Personal Finance
- Corporate Finance
- Insurance
- Taxation
- Budgeting
- Cryptocurrency
- Payment System
- Lending
- Wealth Technology (WealthTech)
- Personal Finance Management
- Insurance Technology (InsurTech)
- Regulation Technology (RegTech)

FinTech Business Challenges:

The major challenges associated with the above FinTech business areas as well as their associated problems, data, and objectives can be categorised into:

- Data Security
- Compliance with Government Regulations
- Lack of Mobile and Tech Expertise
- Big Data and AI Integration
- Blockchain Integration
- User Retention and User Experience
- Effective Marketing Tactics to Acquire Customers
- Personalized Services

USE CASE FOR FINTECH TECHNOLOGIES

In this section, We will discuss different technological challenges and major threats in the finance domain.

Fraud Detection

According to LexisNexis, a business that sells data analytics solutions, financial institutions pay a significant amount for each security lapse, up to three times the money they would have spent to repair the breach. Therefore, artificial intelligence is a lifeline for the banking and financial sectors.

The topic of fraud detection is essential to various businesses, including banking, finance, insurance, and government policies. The main issue challenging the financial and banking industries is the recent increase in digitised fraud. However, because these scams are so rare compared to non-fraud cases, it is difficult for humans to detect fraud from a huge volume of online transaction data.

Earlier, to give a fraud index to any certain payment system, logical arguments and weight summing were used. For example:

-Add 5 points if a seller is 6 months old. Add 2 points if a seller is 12 months old.
-Add 4 points if the transaction amount is twice what it was the previous year. Add 6 points if it happened three times in the last year.
-Similar rules for the size of transactions, mode of payments, *etc.*

Such scores are finally added together, and the magnitude of the sum defines the risk score. As we can see, while this method works, it generates a lot of false positives for new upcoming legitimate vendors and false negatives when a large company commits fraud every now and then.

Different data analysis approaches are utilised today to address problems with the conventional approach. They are separated into two groups:

- Statistical Methods (for Data Preprocessing)
- Machine Learning (for Data Evaluation)

Statistical techniques include creating user profiles and computing multiple average values (*e.g.*, time of call, delay in transaction, *etc.*) These serve as labels for our machine learning algorithms, which will be used to evaluate data [8].

For fraud detection, a supervised machine learning approach is commonly used. We need quality data to use this approach. The features on which the final output is dependent must be present in the data. In fraud detection, it can be vendor names, transaction details such as date, time, location, bank name or source name, and so on. The output or labels have a true or false value to indicate whether the transaction is fraudulent or not. Machine learning models are trained using sample data. After training, the model can be used to accurately predict the future. Detecting fraud requires a classification task. As a result, we can apply a classification model to it. There are numerous machine learning models available for classification tasks. Some of them with promising results are listed below [9 - 10]:

- Logistic regression
- Support Vector Machines
- Neural Networks
- Naive Bayes

Algorithmic Trading

Many financial institutions use algorithmic trading to automate their financial decisions and trades. However, since computers can perform thousands of calculations in a short period of time with extremely high efficiency, it is a very efficient and robust way to conduct financial trading. In algorithmic trading, a mathematical model actively monitors the news and trade results, detecting patterns that can cause stock prices to increase or decline. It can then take proactive action by selling, holding, or buying stocks based on its predictions.

Common machine learning and deep learning models used today include:

- Recurrent Neural Networks(RNN).
- Long Short Term Memory Network(LSTM)
- Ensemble Algorithms
- Support Vector Machines

Machine learning algorithms assist human traders in gaining a marginal advantage over the market average. Given the vast volume of trading operations, even a minor advantage can often translate into substantial profits.

In order to automate financial decision-making processes, machine learning and deep learning techniques have already been widely used in algorithmic trading. Here are some of the principal methods and instances of how they are used in algorithmic trading:

Recurrent neural networks (RNNs): RNNs are a particular class of neural networks that are effective at handling sequential input. RNNs are able to examine time-series data, such as stock prices, in algorithmic trading in order to spot trends and forecast future price alterations. As an illustration, an RNN may be trained on past stock prices to forecast future prices and produce sell or buy signals for algorithmic trading.

Long Short-Term Memory (LSTM) Networks: These RNNs are made to manage long-term dependencies in sequential data. In algorithmic trading, LSTM networks can be employed to examine business news as well as other news-related data in order to spot market trends and forecast how news events would affect stock prices. For instance, to forecast changes in stock values in response to news events, an LSTM may be trained on financial news articles.

Ensemble Algorithms: Ensemble algorithms are machine learning techniques that integrate several different independent models to get a more reliable and accurate overall forecast. Ensemble algorithms can be used in algorithmic trading to integrate various distinct trading techniques, lowering the risk associated with relying just on one approach. For instance, an ensemble algorithm could integrate various stock price prediction models to get a prediction of future stock prices that is more precise and trustworthy.

Support Vector Machines (SVMs): SVMs (Support Vector Machines) are a class of machine learning techniques that may be applied to both classification and regression applications. SVMs can be used in algorithmic trading to categorise financial data into different groups, such as buy, sell, or hold, and to make stock price predictions based on past data. To categorise stocks into several groups and forecast future prices, for instance, an SVM might be trained on historical stock prices.

Note: These are just a few examples of how machine learning and deep learning techniques can be applied in algorithmic trading. The specific methods and applications used depend on the problem being addressed and the data available.

Customer Support

Customer service and recommendation are important factors in the growth of some financial corporations and banks. With the rapidly evolving market, it is critical to advise customers on the best investment strategy, preferably customised recommendations on the basis of their expenditure.

Machine learning assists financial institutions in resolving customer issues quickly by utilising technology such as:

Virtual assistants and chatbots are not only used by these tech firms to improve customer service, but they can also be used in the finance domain. Few tech companies use chatbots to provide excellent customer service. Customers can ask them questions about investments, trends, savings, loans, insurance plans, and so on in order to keep track of their financial situation. Conversations between customers and customer care service executives provide data for virtual assistance and chatbots. It can be a formal or informal conversation between a banking advisor and a customer. Conversations among these two individuals can be fed into neural networks, which will eventually learn to provide appropriate responses to appropriate questions. Recurrent neural networks are widely used for this purpose.

Personalization is essential for increasing customer loyalty and trust in any business or organisation. Machine learning algorithms can analyse customer data and predict which services they might be interested in or provide useful advice. Capital One, for example, has launched the Capital One Second Look programme, which can track spending patterns. After thorough analysis, this programme can detect and notify customers who have been charged twice for the same product or service. Alternatively, it can analyse what tips customers typically leave at restaurants and notify them if they are overly generous. Such programmes make customers feel valued and encourage them to stay with a financial institution.

Sentiment Analysis is the process of analysing customers' emotions, opinions, and attitudes toward other people, products, or services. Machine learning algorithms are used in the financial industry to analyse financial news from various sources and forecast potential stock-market movements. The ability to process massive amounts of data from multiple news channels in seconds is one of the benefits of technology for sentiment analysis.

Risk Management

One of the core concepts of finance is the assessment and balancing of risk. The financial world is full of risk assessment, from online working capital loans to investing choices. DSAIM is the foundation that enables FinTech to build more efficient and accurate credit risk decisions than traditional institutions could ever imagine. DSAIM provides an organisation with in-depth risk analysis and aids risk management by implementing new risk detection system features. By simulating specific risks in a virtual environment, organisations can prepare for potential problems. They can, for example, use risk analysis to determine whether a potential borrower is likely to repay the loan. Risk simulation assists businesses in understanding risk conditions and effectively monitoring them across the organisation. Based on the history of consumer transactions, Data Science creates a risk model. It aids in determining whether users are trustworthy and, if so, whether benefits such as lower rates and additional services can be provided to them [11].

Robotics Process Automation (RPA)

RPA is a software technology that makes it simple to build, deploy, and operate software robots that imitate human actions when engaging with digital systems and software. Software robots, like humans, can understand what is on a screen, complete the correct keystrokes, navigate devices, recognize and retrieve data, and perform a variety of defined actions.

Robotic process automation in financial services operates by following a predetermined workflow that gives robots detailed instructions on what to do at each step. Once robotic process automation software has a roadmap, it can run required programs and perform tasks autonomously as specified by the requirements. RPA automation tools are now used by many banks and financial institutions in opening an account, KYC procedures, anti-money laundering strategies, client request computation, and other recurring tasks.

Blockchain

Blockchain technology creates a secure, decentralised digital ledger that keeps track of all transactions made between computers in a network. Although it was initially developed to power Bitcoin, it has now been used for a number of other uses, notably those in the financial industry.

By removing intermediaries and increasing transaction speed and accuracy, blockchain technology is utilised in fintech to build a more effective, transparent, and secure financial system. For instance, smart contracts that automate intricate financial transactions can be made using blockchain technology, as well as decentralised payment systems, digital assets, and currencies.

Additionally, by offering a permanent, impenetrable record of transactions, lowering the risk of fraud, and enhancing transparency, blockchain technology supports the development of a more secure and transparent financial environment.

In general, the adoption of blockchain technology in fintech is reshaping the financial sector and presenting fresh chances for innovation and expansion [12, 13].

Application of Blockchain in FinTech:

The fintech sector has many uses for blockchain technology, including:

- Digital payments: Compared to traditional payment methods, blockchain-based payment systems can offer faster, less expensive, and more secure transactions.
- Digital Assets and Cryptocurrencies: Blockchain makes it possible to create and manage digital assets and cryptocurrencies, which facilitates ownership tracking, trading, and transfer.
- Supply Chain Finance: Blockchain can be used in supply chain finance to boost efficiency, cut costs, and promote transparency.
- Decentralised Exchanges: Blockchain can be used to build decentralised exchanges, allowing for peer-to-peer trade without the involvement of middlemen.
- Identity Management: Blockchain-based identity management systems can offer efficient and secure solutions to store, manage, and utilise data related to a person's identification.
- Crowdfunding: Blockchain technology can be used to build decentralised platforms for crowdfunding, allowing businesses and individuals to raise money

from investors directly and without the use of middlemen.
- Insurance: By utilising blockchain technology, decentralised insurance platforms can be built, providing real-time claims processing, fraud detection, and cost savings over conventional insurance models.

These are a few of the most well-known blockchain applications in the finance industry, and the list is continually expanding as more companies learn about the potential advantages of this technology [14 - 20].

The following are a few of the most widespread applications of blockchain technology in the fintech sector:

- Cross-border payments: These are possible almost instantly with fewer fees and more transparency thanks to blockchain-based payment systems.
- Blockchain technology can be used to develop decentralised systems for peer-to-peer lending that connect lenders and borrowers directly, eliminating the need for a middleman.
- Remittances: Blockchain-based remittances can offer quicker, less expensive, and safer options for transmitting money across international borders.
- Digital Identification: Blockchain-based solutions for digital identity can offer secure, decentralised, and user-controlled ways to store and use data about a person's identity.
- Decentralised systems for investment management can be created using blockchain technology, providing investors with more security and transparency.
- Trade Finance: Blockchain-based platforms for trade finance can automate and streamline transactions, lowering the possibility of fraud and boosting productivity [21 - 29].
- Digital Know-Your-Customer (KYC): Blockchain technology can be used to develop decentralised digital Know-Your-Customer (KYC) solutions, which save costs and improve the effectiveness of onboarding new customers [30 - 32].

These use cases demonstrate the wide range of applications for blockchain technology in the fintech industry, from improving existing financial services to creating new and innovative solutions.

CONCLUDING REMARKS

FinTech is a new development, and there are few studies on its socioeconomic, governmental, technical, and organisational aspects. This makes it difficult for financial institutions to make well-informed decisions about investing in fintech projects. This paper discussed fintech's areas of business, corresponding challenges, the use of DSAIM for the mentioned problems, as well as some use cases where DSAIM assists these fintech businesses. Overall, these emerging technologies have the potential to greatly improve the efficiency, security and accessibility of financial services. It is important for fintech companies and financial institutions to stay up-to-date with these advancements and consider how they can be applied to their business.

REFERENCES

[1] L. Cao, Q. Yang, and P.S. Yu, "Data science and AI in FinTech: An overview", *Int. J. Data Sci. Anal.,* vol. 12, no. 2, pp. 81-99, 2021. [http://dx.doi.org/10.1007/s41060-021-00278-w]

[2] I. Lee, and Y.J. Shin, "Fintech: Ecosystem, business models, investment decisions, and challenges", *Business Horizons,* vol. 61, no. 1, pp. 35-46, 2018. [http://dx.doi.org/10.1016/j.bushor.2017.09.003]

[3] G. Suseendran, E. Chandrasekaran, D. Akila, and A.S. Kumar, *Banking and FinTech (Financial Technology).* Embraced with IoT Device, 2019. [http://dx.doi.org/10.1007/978-981-32-9949-8_15]

[4] M. Javaid, A. Haleem, R.P. Singh, R. Suman, and S. Khan, "A review of blockchain technology applications for financial services", *BenchCouncil Transactions on Benchmarks, Standards and Evaluations,* vol. 2, no. 3, p. 100073, 2022. [http://dx.doi.org/10.1016/j.tbench.2022.100073]

[5] "Fintech empowerment: Data science, AI, and machine learning", Available from: https://www.cutter.com/article/fintech-empowerment-data-science-ai-and-machine-learning-501881

[6] L. Cao, G. Yuan, T. Leung, and W. Zhang, "Special issue on AI and FinTech: The challenge ahead", *IEEE Intell. Syst.,* vol. 35, no. 2, pp. 3-6, 2020. [http://dx.doi.org/10.1109/MIS.2020.2983494] [PMID: 35935446]

[7] G. Pisoni, B. Molnár, and Á. Tarcsi, "Data science for finance: Best-suited methods and enterprise architectures", *Appl. Syst. Innov.,* vol. 4, no. 3, p. 69, 2021. [http://dx.doi.org/10.3390/asi4030069]

[8] F. Lacruz, and J. Saniie, "Applications of machine learning in fintech credit card fraud detection", *2021 IEEE International Conference on Electro Information Technology (EIT).* 14-15 May 2021, Mt. Pleasant, MI, USA, 2015. [http://dx.doi.org/10.1109/EIT51626.2021.9491903]

[9] D.P.K. Priya, and D.K. Anusha, "Fintech issues and challenges in India", *Int J Recent Technol Eng,* vol. 8, no. 3, pp. 904-908, 2019. [http://dx.doi.org/10.35940/ijrte.C4087.098319]

[10] S. Fernandez-Vazquez, R. Rosillo, D. De La Fuente, and P. Priore, "Blockchain in FinTech: A mapping study", *Sustainability,* vol. 11, no. 22, p. 6366, 2019. [http://dx.doi.org/10.3390/su11226366]

[11] T. Renduchintala, H. Alfauri, Z. Yang, R.D. Pietro, and R. Jain, "A survey of blockchain applications in the FinTech sector", *J. Open Innov.,* vol. 8, no. 4, p. 185, 2022. [http://dx.doi.org/10.3390/joitmc8040185]

[12] F. Lacruz, and J. Saniie, "Applications of machine learning in fintech credit card fraud detection", *2021 IEEE International Conference on Electro Information Technology (EIT),* 2021. 14-15 May 2021, Mt. Pleasant, MI, USA

[13] S.K. Das, and N. Dey, *Constraint Decision-Making Systems in Engineering* IGI Global, 2023.

[14] S. Majumder, and N. Dey, *AI-empowered knowledge management.* Springer, 2022. [http://dx.doi.org/10.1007/978-981-19-0316-8]

[15] S.K. Das, "Smart design and its applications: Challenges and techniques", In: *Nature-Inspired Computing for Smart Application Design,* S.K. Das, TP. Dao, T. Perumal, Eds., Springer: Singapore, 2021.

[16] S.K. Das, N. Dey, G.R. Crespo, and H.E. Viedma, "A non-linear multi-objective technique for hybrid peer-to-peer communication", *Inf. Sci.,* vol. 629, pp. 413-439, 2023. [http://dx.doi.org/10.1016/j.ins.2023.01.117]

[17] A.K. Dass, and S.K. Das, An optimal data communication system in wireless sensor network using artificial neural network.*Constraint Decision-Making Systems in Engineering.* IGI Global, 2023, pp. 120-134. [http://dx.doi.org/10.4018/978-1-6684-7343-6.ch007]

[18] N. Dey, Ed., *Data-Driven Approach for Bio-medical and Healthcare.* Springer Nature, 2022.

[19] A. Sarker, "Error detection architectures for hardware/software co-design approaches of number-theoretic transform", *IEEE Transactions on Computer-Aided Design of Integrated Circuits and Systems,* pp. 2418-2422, 2023.

[20] M.M. Kermani, "Integrating emerging cryptographic engineering research and security education", *2015 ASEE Annual Conference & Exposition,* pp. 26-289, 2015.

[21] M. Mozaffari-Kermani, Reliable and high-performance hardware architectures for the Advanced Encryption Standard/Galois Counter Mode Doctoral dissertation, The University of Western Ontario Canada, 2011.

[22] A. Sarker, and M. M. Kermani, "Efficient error detection architectures for postquantum signature falcon's sampler and KEM SABER", *IEEE Transactions on Very Large Scale Integration (VLSI) Systems,* vol. 30, no. 6, pp. 794-802, 2022.

[23] A.C. Canto, J. Kaur, and M.M. Kermani, "Algorithmic security is insufficient: A comprehensive survey on implementation attacks haunting post-quantum security", *arXiv:2305.13544,* 2023.

[24] R. Azarderakhsh, B. Koziel, and S. F. Langroudi, "FPGA-SIDH: High-performance implementation of supersingular isogeny Diffie-Hellman key-exchange protocol on FPGA", *IACR Cryptol. ePrint Arch.,* vol. 2016, p. 672, 2016.

[25] E. Dubrova, K. Ngo, and J. Gärtner, "Breaking a fifth-order masked implementation of crystals-kyber by copy-paste", *Proceedings of the 10th ACM Asia Public-Key Cryptography Workshop,* pp. 10-20, 2023. [http://dx.doi.org/10.1145/3591866.3593072]

[26] J. Kaur, A.C. Canto, and M.M Kermani, "A comprehensive survey on the implementations, attacks, and countermeasures of the current NIST lightweight cryptography standard", *arXiv preprint arXiv:2304.06222,* 2023.

[27] M. M. Kermani, and R. Azarderakhsh, "Error detection reliable architectures of Camellia block cipher applicable to different variants of its substitution boxes", *In 2016 IEEE Asian Hardware-Oriented Security and Trust (AsianHOST),* pp. 1-6, 2016.

[28] C.A. Canto, J. Kaur, and M.M Kermani, "ChatGPT vs. Lightweight Security: First Work Implementing the NIST Cryptographic Standard ASCON", *arXiv preprint arXiv:2306.08178,* 2023.

[29] M.M. Kermani, E. Savas, and S.J. Upadhyaya, "Guest editorial: Introduction to the special issue on emerging security trends for deeply-embedded computing systems", *IEEE Trans. Emerg. Top. Comput.,* vol. 4, no. 3, pp. 318-320, 2016. [http://dx.doi.org/10.1109/TETC.2015.2482778]

CHAPTER 7

Leveraging Blockchain Technologies in Healthcare Applications

Manorama Patnaik[1,*]

[1] *Amity Institute of Information Technology, Amity University, Jharkhand, Ranchi, India*

Abstract: Blockchain innovation recently led to key innovations in the advanced insurgency of medical care, but a few studies have shown that blockchain has the potential to improve the medical care environment. It is prepared to change how customary clinical frameworks and organizations have been occupied with the medical services area throughout the previous very long while. Information and communication technologies (ICTs) and blockchains are key empowering advancements for the decentralization and digitalization of medical care foundations and provide current and digitalized medical care environments to patients, similar to specialist organizations. With regard to blockchain applications for medical service information, the board provides utility for patients, specialists and medical care organizations through patient record access and control, cases and installments, clinical IoT security and exploration of information checks and trades for budgetary evaluation and straightforwardness. In these applications, constant updates to an encoded, decentralized blockchain record are never true, screen, or control clinical data. This likewise encourages medical service foundations to confine unapproved individuals to obtaining sensitive data. This paper provides insight into the blockchain methodology applied in IoT healthcare security.

Keywords: Blockchain technology as supply chain management, Decentralization, Electronic health records, Healthcare concerns in technology management, Internet of medical things (IoMT), Supply chain management.

INTRODUCTION

A blockchain is a dispersed public ledger information/database that is kept up by an organization of checked members or hubs and stores unchanging information that can be shared safely without outside intervention [1]. Information is saved and recorded with cryptographic marks and through the use of agreement calculations that are established as key empowering agents for its application. This capacity for information conservation is a major explanation that has driven

* **Corresponding author Manorama Patnaik:** Amity Institute of Information Technology, Amity University, Jharkhand, Ranchi, India; E-mail: Manorama7826@gmail.com

the utilization of blockchain in medical services, wherein a large amount of information is dependent upon broad trade and conveyance [2]. The development of blockchain innovation and its application in different settings has taken place in different stages [3]. The principal period of blockchain advancement was identified with digital currency, and the second was related to the use of smart agreements in regions, for example, land and funds [4]. The third era of development was centered around the utilization of blockchain in non-financial areas, for example, the government, medical care, and culture. Additionally, determined by creative, innovative highlights, for example, an immutable nature, blockchain is currently viewed as being in its fourth phase of advancement through the fusion of computerized reasoning (AI). The variety of blockchains utilized might be ascribed to its potential for establishing decentralized and trustless exchange conditions [5]. The IoT healthcare industry is a prime contender for blockchain technology, as blockchain can possibly address basic concerns, for example, automated guarantee approval and the general well-being of executives. This innovation may permit patients to claim information and pick with whom it is shared, consequently tending to deal with problems of information proprietorship and sharing. Simultaneously, information records can be bound together, refreshed, safely traded, and accessed opportunistically by suitable specialists *via* the use of agreement conventions. This is a significant favorable position managed by the utilization of blockchain in medical services since current practices expect information to be put away with outsiders for authentication purposes. Because authorized person can only access the medical data when required [6]. Hence, blockchain can directly adopt the technologies to carry straightforwardness to information about broad measures while additionally decreasing the odds of information misuse or abuse due to conceivable human mistakes. Despite the positive implications of blockchain for cultural and business change, there is a discussion on its pervasive favorable circumstances and determined advantages in contrast to recently met requirements. It indicates the requirements of the customers based on real-life applications that are evolving day by day [7]. A report suggests that while organizations may pursue blockchain-based advancements, they may adopt a more simple approach due to concerns about potential advertising excessively. It might be said that this innovation is yet to live up to its promoted desires, a reality that might be ascribed to specific difficulties in the far-reaching usage of this innovation, particularly regarding administrative boundaries [8]. Another significant test in proclaiming the organization of blockchain is the newness of the general population and individual clients, for example, patients or specialists, given the way this innovation works, its specialized highlights or its advantages for healthcare administration [8].

This chapter provides deep insight into healthcare-based privacy adoption using blockchain technologies. The introduction section represents the adoption of blockchain technologies in healthcare and healthcare-assisting technologies. Section 2 discusses the types of blockchains. Section 3 defines healthcare-based blockchain applications. Section 4 describes a case study based on Ethereum. Section 5 describes the challenges faced by adopting blockchain technologies. The conclusion and future scope are explained in Section 6 and Section 7, respectively.

TYPES OF BLOCKCHAIN

Bitcoin is a digital currency that uses blockchain as its foundation. By upholding its fundamental character traits of decentralization, anonymity, permanence, and suitability for the e-cash exchange measure, it has elevated its sheer relevance in the digital world. There are four different types of blockchains [9].

Public Blockchains

Public blockchains are decentralized, open-source blockchains that are publicly accessible and have no impact on the number of users who might express interest in the project. Since there is no structure for setting up exchanges, anyone can join and read, write, or evaluate the blockchain without the organization being controlled by a single company [10]. According to its objective, public blockchains are the best at ensuring client anonymity. Since this form of blockchain is open to all users, judgments are made *via* a variety of agreement calculations, such as proof of work (POW) and proof of stake (POS) [10]. Additionally, the public blockchain platform retains an incentive component predefined in the convention through some gaming hypothesis, meaning that the group's members are paid monetarily to maintain the framework's best practices and dependability [11]. The public blockchain stage includes Bitcoin, Ethereum, Litecoin, and so on.

Private Blockchains

Private blockchain restricts clients who show interest in the business and conduct transactions there. The blockchain network is under the management of a group of individuals or organizations that are permitted entry into the organization. As a result, a private blockchain has a client personality from the very beginning for choosing its specific tasks inside the organization and managing access, such as reading, creating, or examining explicit data in the blockchain. The plan of the private blockchain, in contrast to the public blockchain, is more incorporated, so the choices are made by an in-control who allots a few rights to the members of the organization [10]. Nonetheless, the incorporated engineering of a private blockchain increases the inclination for security penetration. The private blockchain is utilized by associations that require versatility, information assurance security, and administrative standards for state consistency [11]. In this way, particular members who are predefined with explicit rules in the organization approach the block of data from inside, confirming and approving the exchanges. The private blockchain stages include Hyperledger, Hashgraph, Corda, and numerous others.

Consortium/Federated Blockchains

A consortium blockchain is a partially decentralized blockchain that is present between a public and a private blockchain, as the name suggests. Although the consortium blockchain only partially demonstrates the characteristics of both public and private blockchains, it contains a great majority of private blockchain characteristics. In contrast to a private blockchain, a consortium blockchain is organized by a group of components [12]. A consortium blockchain, in contrast to a public blockchain, does not enable anyone to join the network; instead, the organization's administrators need to approve any new member. The authority of the organization is granted in advance to a specified hub reliant on this type of blockchain, which is frequently referred to as “semi-private”.

This sort of blockchain is regularly referred to as a “semi-private” blockchain since the authority of the organization is given ahead of time to a specifically predefined set of hubs dependent on a few agreement calculations. For the most part, it is utilized in business associations with few colleagues. Instances of consortium stages include R3 and Corda.

Hybrid Blockchains

A hybrid blockchain is a blend of both public and private blockchains. It joins the elements of interest of each blockchain; in other words, a hybrid blockchain represses the security advantages of a private blockchain and the straightforwardness advantages of a public blockchain as needed. The licensed interchain capacity offered ascends to the hybrid nature of the blockchain, empowering the hybrid blockchain to have numerous chain organizations of blockchains [13]. The hybrid blockchain is not completely open to everyone; certain limitations are posted while permitting members to enter the hybrid blockchain network. This hybrid nature of this stage gives associations the appropriately controlled admittance of their information to the organization along these lines, making the framework more adaptable and secure without bargaining for protection. Despite the fact that the hybrid blockchain is constrained by a gathering of people, the exchanges made are kept hidden but can be checked at any point needed. This approach maintains the immutability of the exchanges [14]. The hybrid blockchain incorporates Dragonchain technology.

BLOCKCHAIN-BASED IOT HEALTHCARE APPLICATIONS

The information demonstrated and administered in many medical care applications are reclassified by blockchain technology. This is due to its adaptability and exceptional ability to portion, safeguard, and exchange clinical information and services. Many recent advancements in the medical services sector have focused on blockchain technology. The development of blockchain-based medical service innovations can be roughly divided into four layers: partners, IoT medical data management apps, data sources for supply chain management, and blockchain innovation in the Internet of Medical Things. Fig. (**1**) depicts a blockchain-based work process for medical service industry applications.

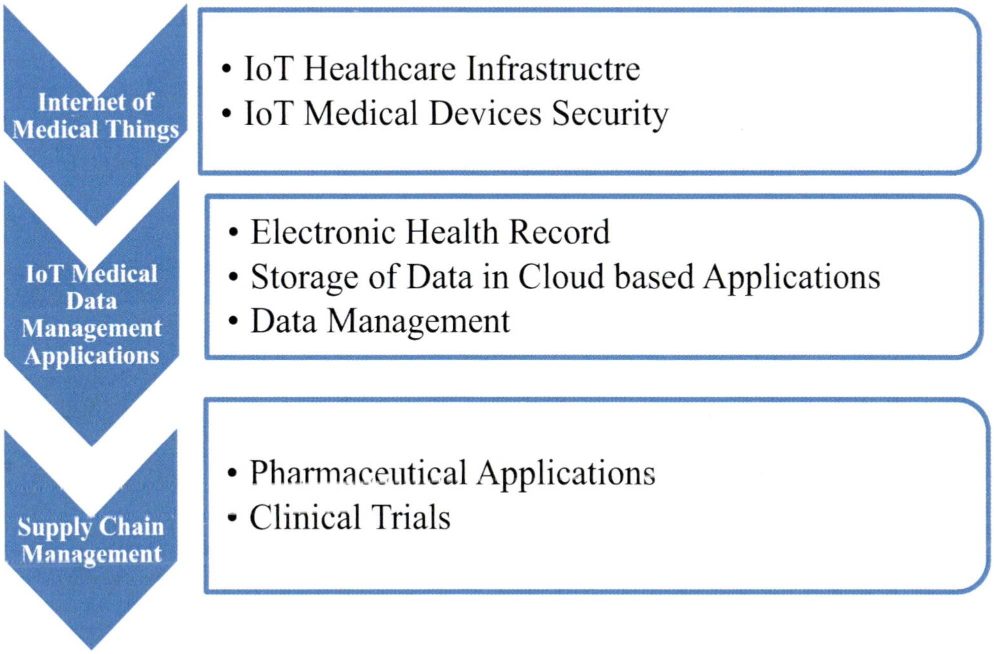

Fig. (1). Blockchain-based healthcare applications.

Initially, all the data from scientific instruments, laboratories, online media, and countless other sources were merged to create basic data that were then scaled up to massive data. The primary layer of the stack comprises this information, which is the key component of the overall blockchain-based medical care system. Blockchain technology is positioned on top of the basic information layer and is thought of as the main support structure for a medical services design separate into four parts. The numerous highlights of each blockchain stage include agreement calculations and conventions [15]. The blockchain stages encourage clients to make and address their exchanges. A few blockchain stages were created and are presently being used; for example, Ethereum, Ripple, and Hyperledger Fabric [16 - 18]. The essential parts of the blockchain are shrewd contracts, marks, wallets, occasions, enrollment and advanced resources. A wide range of conventions could be used to communicate with other projects and hierarchies or even across different corporations. For instance, P2P, concentrated, decentralized, and disseminated may be included. Depending on the range of needs they must meet, policymakers may decide to take an open, private, or even unified position. The next step is to ensure that the applications are compatible with the complete framework after the process is created by implementing blockchain technology. The applications of healthcare *via* blockchain can be categorized into three broad categories: Information stockpiling (using cloud-based apps, for example), global logical information exchange for research and development (R&D), and electronic health records (EHRs) [19]. The second-rate course discusses IoMT, including the conversion of medical IoT and clinical devices, the medical IoT framework and information security, and AI. The applications for managing healthcare data on a blockchain are shown in Fig. (**2**). The partner layer, which includes groups that profit from blockchain-based medical care apps, such as business clients, specialists, and patients, appears last and at the top of the sequence. In this layer, clients' primary concerns are how to monitor, distribute, and manage information effectively without jeopardizing security and protection.

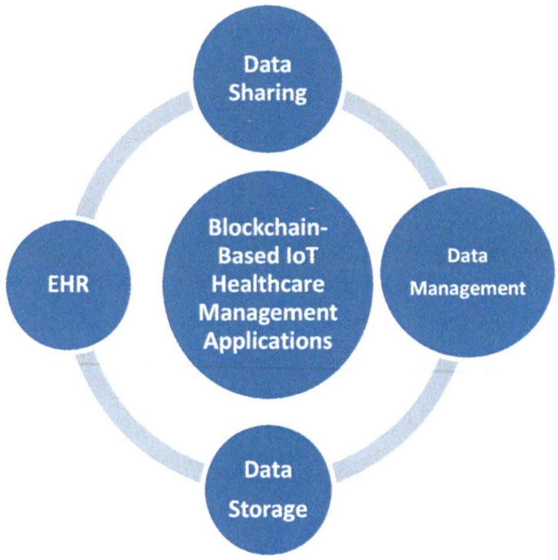

Fig. (2). Components of blockchain-based IoT healthcare management applications.

Electronic Health Records (EHRs)

EHRs were introduced to replace traditional clinical records as a result of data innovation. Doctor practices were able to significantly improve the nature of treatment thanks to electronic access to health records [20]. Moreover, EHR enables improved illness management and increased levels of preventive consideration. Additionally, the updated record offers extended parental cooperation and improved choice assistance. In keeping with this, the medical services network is increasingly acknowledging its role [21]. Several studies have been conducted to develop blockchain technology for the purpose of securing, providing, and storing EHR data both within and across companies [22]. In this structure, clinical information is obtained from executives utilizing a computerized file that approaches the control privileges of their proprietors' data. This is accomplished by conveying cloud encryption under the chain. In a connected report, Guo *et al.* presented a property-based mark conspire using blockchain innovation with different specialists to ensure and approve EHRs [23]. This approach encourages a bunch of messages to communicate and oppose agreement assaults. Wang and Song proposed a protected cloud-based EHR framework in light of blockchain and characteristic-based cryptosystems [24]. To scramble clinical information, they utilized a mix of personality-based encryption and character-based marks simultaneously to actualize advanced signs. In the head of that blockchain, different methods are utilized to guarantee the uprightness and recognizability of the clinic. While the previously mentioned three investigations have principally centered around cryptographic angles to secure EHR blocks, Roehrs *et al.* have tended to identify various activities through the unification of dispersed well-being records and the inclusion of executives from medical care providers' partners [25]. These two issues were comprehended by proposing OmniPHR, an appropriate model for incorporating Physical health records (PHRs) that utilizes an equal information base to store PHRs in blocks and consolidated auxiliary semantic interoperability and cutting-edge vision of various PHR designs. Finally, using an extraordinary approach, Hussein *et al.* constructed a system for ensuring clinical records utilizing blockchain innovation dependent on hereditary calculations and discrete wavelet changes [26].

Data Sharing

The sharing of medical services and clinical information is a fundamental and basic advancement in improving the quality of medical service suppliers and medical services framework [27 - 29]. Sharing well-being records occur between people; for example, a patient who needs to provide his or her clinical history to a specialist at their first gathering [30]. Additionally, sharing can occur between an individual and a partner, for example, a patient imparting his clinical history to an insurance agency. Indeed, this information can even be shared past the outskirts [31]. However, the operational system of the present wellbeing-related frameworks has a few limitations. One restriction is that patients scarcely admit to their well-being records. In this manner, they have no clue about sharing their own well-being information among obscure gatherings [32]. To improve communication and coordinated effort with the medical care industry, blockchain innovation could assume a critical job, empowering and ensuring the sharing of advantageous electronic well-being information. This approach is viewed as one of the most critical commitments of blockchain-based medical services [33].

Data Storage

On a decentralized storage system, every transaction in a blockchain based on the healthcare industry is saved in blocks. EHRs, which are regarded as the cornerstones of sizable distributed medical storage, are used in the healthcare system to arrange the medical data of patients [34]. Where security is a top priority, this can be kept on-site or in the cloud. To support a large amount of information technology infrastructure, cloud storage is composed of several large volumes of connected storage devices. One illustration of such an IT infrastructure is a healthcare system built on a blockchain. Fast transmission, good sharing, large storage capacity, low cost, simple access, and dynamic association are all benefits of cloud storage technology. As proposed by Al Omar *et al.*, blockchain technology has been recommended as a storage method for a cloud-based patient-centric healthcare data management system that aids in preserving anonymity [35]. By creating a set of security and privacy requirements to accomplish accountability, integrity, and security, the major goal of this effort is to maintain private healthcare data on the blockchain. Kaur *et al.* coined the term "BlockCloud" to describe the combination of blockchain technology applied in a cloud setting [36]. Using the cloud is intended to ensure that the data are disseminated and secure under one roof without involving other parties. The study addressed difficulties in creating and enforcing policies through collaboration between healthcare service providers, public health groups, and governments [37].

Data Management

Even though many organizations, especially medical service foundations, are information-driven and the amount of information produced in this era or another, such as the IoT, is constantly growing, information security and security are constantly violated, either intentionally or unintentionally, by both legitimate and illicit users [38 - 40]. As a result, a great deal of money and reputation have been lost by countless organizations. Different clients of health information have different jobs; thus, access to information should be reflected by the benefits assigned to these jobs. This method of access can be guaranteed, in a consistent way, by blockchain innovation. We depict a few blockchain innovations that have been produced for such a reason. MedRec is a decentralized EMR framework in which information authorization and activity are recorded in the blockchain, and execution is completed through brilliant agreements [41]. The MedRec team takes up the provider's finished clinical data for information verification, privacy, examination, and sharing and gives patients extensive, permanent clinical information and administration. In another examination, Zhu *et al.* proposed a methodology for obtaining controllable blockchain information from executives in a cloud climate to address the worries of clients about the absence of control of posted records [39]. In their model, they planned a unique trust authority hub to permit clients to end and forestall any possibly malignant activities, even in a greater part, assault. In another commitment made by Genestier *et al.*, the groundbreaking idea of reshaping the assent of the board in the medical services framework, which, for the most part, allows clients to control all of their well-being records by utilizing blockchain, was presented [40]. Nonetheless, there is no approval plan and no entrance control during their execution.

CASE STUDY: ETHEREUM-BASED SMART CONTRACT SYSTEM

Clinical analysts, healthcare providers, and individuals all benefit from blockchain innovation in different ways [42]. Making a single repository for all health information, tracking customized information continuously, and granularly setting information access consent forms would serve research just as effectively as personalized medicine [43]. To advance the understanding of infection, accelerate biomedical discovery, track the improvement of pharmaceuticals quickly, and develop customized treatment plans based on genetic characteristics, life operations, and climate, health analysts require extensive informational collections [44]. The shared information structure of blockchain provides a broad spectrum of informative indices by including patients from a variety of racial, economic, and geographic backgrounds [45]. Since blockchain collects data, it provides the best data for longitudinal studies on well-being information over the lifetime of an individual [46]. A healthcare blockchain increase the variety of health information available by incorporating data from groups of people who are now underserved by the clinical network or who are not typically interested in research [47]. The widespread information environment of blockchain systems makes it easier for "hard-to-reach" masses to be intrigued and for the general populace to produce more intelligent results [48, 49]. Another class of "intelligent" medical service provider applications that would avoid the most recent clinical studies and produce modified treatment routes would likely be made possible by a medical services blockchain [50, 51]. The provider of medical services and the patient approach the same facts and may participate in an amicable, educated discourse about exploration-based therapy options rather than instinct-based best-case scenarios [52].

A decentralized application (DApp) that supports a private blockchain network and a back-end distributed file system implements the framework (DFS). The blockchain smart contract system for healthcare has been implemented using Ethereum. With a vibrant community and a sizable public DApp repository, this open-source network is currently one of the largest public blockchain networks. The platform currently uses the Ethash proof-of-work (PoW) consensus algorithm, but developers are hoping to switch to a proof-of-stake (PoS) scalability algorithm soon. Distributed applications should ideally be designed using a consensus technique that uses delegated proof-of-stake (DPoS) or practical byzantine fault tolerance (PBFT). The DApp able to identify abnormalities, unauthorized data insertions, and missing entities by comparing DFS content with ledger records. There is a timeline for auditing after each stage. The fundamental components of smart contracts are functions, events, state variables, and modifiers, and they are created using the high-level programming language known as Solidity. The Rect and Kovan test networks have been used to pay transaction fees for testnet ethers and smart contracts. Writing, compiling, and announcing are the three steps in the construction of a smart contract using solidity programming [53]. Solidity's real-time compiler creates the bytecode. Ethereum Wallet has been used to publish smart contracts to the blockchain. Fig. (**3**) shows how smart contracts work with Ethereum, excluding the mining process, for ease of understanding.

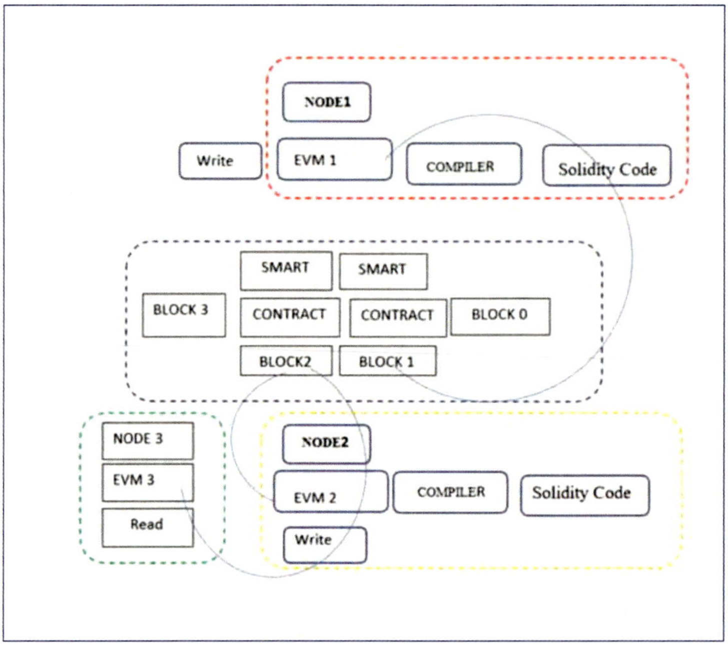

Fig. (3). Ethereum-based smart contract method.

This smart contract is machine-compiled into byte code, where each byte represents an operation and is then uploaded to the blockchain as an EVM-1 transaction. A miner picks it up and confirms Block 1. The EVM-2 retrieves the web-based data, embeds it in Transaction tx, and deploys it to the blockchain after a user delivers the request through the web interface. Block 2 updates the transaction tx's status. Node 3 must synchronize up to at least Block 2 to notice the changes that tx causes when it subsequently needs to check the states that are stored in the contract. Fig. (**3**) illustrates the aforementioned smart contract framework.

DRAWBACKS OF BLOCKCHAIN IN THE INTERNET OF MEDICAL THINGS

- **Absence of Standardization**

The medical care industry is receiving best-in-class IoMT innovation. Notwithstanding medical service suppliers on the premise of clinical devices, the IoMT opens the office of clinical administrations to be able to universally and perpetually direct patients. This is similar to the individual block of the telehealth observation applications, as remote devices are related to catching biosignals. Accordingly, carefully designed transmission of clinical information is important [54]. This issue results from the high heterogeneity in correspondence conventions, clinical gadgets and stages, which could be attributed to the absence of normalization among IoMT markers. In this manner, securing such a factor-normalized transmission medium is troublesome from the current viewpoint [55]. Cutting-edge innovation, for example, blockchain, could be applied to stand up to unapproved access during information transmission within an obscure organization, as the differing correspondence standards of a framework do not influence blockchain's exhibition.

- **Lack of Adoption of Cryptocurrencies**

The majority of medical professionals do not accept cryptocurrencies for payment. Online payments were not widely used prior to the adoption of blockchain technology. There is still a cash flow. Hence, correct online payment implementation should be used. Thus, blockchain technology should be used.

- **Absence of Technical Expertise**

Not all users are anticipated to have expensive hardware and software resources. Many users are likewise uninitiated current technological developments. For instance, many senior citizens avoid using laptops or PCs. Cryptocurrency mining requires GPUs, which are not found in all laptops. This is a significant problem for blockchain.

- **Lack of Acceptance of Paperless Methods**

Paper records are preferred by many consumers and clinicians. They favor using a file system to organize medical records. Not all pharmacies are completely paperless. Prescriptions are typically used by pharmacies to maintain track of their inventory. Patients save paperwork for their own convenience as well. Therefore, adjusting to a blockchain network that is completely paperless is a difficult undertaking.

CONCLUSION

The medical care sector is moving much more quickly toward receiving best-in-class IoMT innovations. In addition to medical service providers with on-site clinical equipment, IoMT offers the office of clinical administrations that can be accessed globally to guide patients continuously. Moreover, telehealth observation applications are structured because they use distant devices to collect biosignals. As a result, proper clinical information transmission is important [54]. Due to the significant degree of variety across IoMT manufacturers' communication conventions, clinical devices, and phases, this problem arises. Several cases, such as charging and installing executives, were introduced in this paper as future examination issues. Finally, we introduced the utilization of blockchain in IoMT.

FUTURE SCOPE

The medical services frameworks of the 21st century will comprise different gadgets associated with patients (*e.g.*, far-off medical care administrations, wearable gadgets, and so forth) and their guardians. These frameworks produce information ceaselessly and can be dependent on vindictive assaults while being in transmission at different degrees of hidden correspondence organization. In this chapter, we examined a few studies that propose alternative safe frameworks utilizing blockchain innovation to guarantee the confidentiality of well-being information. The primary issue that needs consideration from specialists is the manner in which the blockchain works in unpredictable and various correspondence frameworks. The IoMT conveyance framework utilizes correspondence networks constructed by various assistance suppliers with various information access control arrangements.

REFERENCES

[1] J. Michael, A. Cohn, and J.R. Butcher, "BlockChain Technology", Available from: https://www.steptoe. com/images/content/1/7/v3/171269/LIT-FebMar18-Feature-Blockchain.pdf

[2] B. Bhushan, and N. Sharma, "Transaction Privacy Preservations for Blockchain Technology", *Advances in Intelligent Systems and Computing International Conference on Innovative Computing and Communications,* pp. 377-393, 2020. [http://dx.doi.org/10.1007/978-981-15-5148-2_34]

[3] S. Soni, and B. Bhushan, "Transaction Privacy Preservations for Blockchain Technology", *Advances in Intelligent Systems and Computing International Conference on Innovative Computing and Communications,* pp. 377-393, 2019. [http://dx.doi.org/10.1109/ICICICT46008.2019.8993210]

[4] K. Yaeger, M. Martini, J. Rasouli, and A. Costa, "Emerging blockchain technology solutions for modern healthcare infrastructure", *J. Scient. Innov. Med.,* vol. 2, no. 1, p. 1, 2019. [http://dx.doi.org/10.29024/jsim.7]

[5] W.J. Gordon, and C. Catalini, "Blockchain technology for healthcare: facilitating the transition to patient-driven interoperability", *Comput. Struct. Biotechnol. J.,* vol. 16, pp. 224-230, 2018. [http://dx.doi.org/10.1016/j.csbj.2018.06.003] [PMID: 30069284]

[6] T.K. Mackey, T.T. Kuo, B. Gummadi, K.A. Clauson, G. Church, D. Grishin, K. Obbad, R. Barkovich, and M. Palombini, "'Fit-for-purpose?' challenges and opportunities for applications of blockchain technology in the future of healthcare", *BMC Med.,* vol. 17, no. 1, p. 68, 2019. [http://dx.doi.org/10.1186/s12916-019-1296-7] [PMID: 30914045]

[7] M.A. Engelhardt, "Hitching healthcare to the chain: An introduction to blockchain technology in the healthcare sector. Technology Innovation Management Review, 7(10).4. Yaeger, K.; Martini, M.; Rasouli, J.; Costa, A. Emerging Blockchain Technology Solutions for Modern Healthcare Infrastructure", *J. Sci. Innov. Med.,* vol. 2019, p. 2, 2017.

[8] T.M. Fernández-Caramés, and P. Fraga-Lamas, "A Review on the Use of Blockchain for the Internet of Things", *IEEE Access,* vol. 6, pp. 32979-33001, 2018. [http://dx.doi.org/10.1109/ACCESS.2018.2842685]

[9] X. Li, P. Jiang, T. Chen, X. Luo, and Q. Wen, "A survey on the security of blockchain systems", *Future Gener. Comput. Syst.,* vol. 107, pp. 841-853, 2020. [http://dx.doi.org/10.1016/j.future.2017.08.020]

[10] S. Khatwani, "Different Types Of Blockchains In The Market and Why We Need Them", https://coinsutra. com/different-types-blockchains/(13.3.2019.)

[11] S.G. Sharma, L. Ahuja, and D.P. Goyal, "Building secure infrastructure for cloud computing using blockchain", *2018 Second International Conference on Intelligent Computing and Control Systems (ICICCS),* IEEE., pp. 1985-1988, 2018. [http://dx.doi.org/10.1109/ICCONS.2018.8663145]

[12] A. B. Haque, and M. Rahman, "Blockchain Technology: Methodology, Application and Security Issues", *IJCSNS,* vol. 20, no. 2, 2020.

[13] S. Gupta, S. Sinha, and B. Bhushan, "Emergence of Blockchain Technology: Fundamentals, Working and its Various Implementations", *SSRN,* 2020. [http://dx.doi.org/10.2139/ssrn.3569577]

[14] P. Zhang, D.C. Schmidt, J. White, and G. Lenz, "Blockchain technology use cases in healthcare", *Adv. Comput.,* vol. 111, pp. 1-41, 2018. [http://dx.doi.org/10.1016/bs.adcom.2018.03.006]

[15] C. Saraf, and S. Sabadra, "Blockchain platforms: A compendium", *2018 IEEE International Conference on Innovative Research and Development (ICIRD),* pp. 1-6, 2018.

[16] S. Bistarelli, G. Mazzante, M. Micheletti, L. Mostarda, and F. Tiezzi, "Analysis of ethereum smart contracts and opcodes", *International Conference on Advanced Information Networking and Applications,* Springer: Cham, pp. 546-558, 2019.

[17] J. Göttfert, *Cointegration among cryptocurrencies: A cointegration analysis of Bitcoin.* Bitcoin Cash, EOS, Ethereum, Litecoin and Ripple, 2019.

[18] Z. Shi, H. Zhou, Y. Hu, S. Jayachander, C. de Laat, and Z. Zhao, "Operating permissioned blockchain in clouds: A performance study of hyperledger sawtooth", *2019 18th International Symposium on Parallel and Distributed Computing (ISPDC),* pp. 50-57, 2019. [http://dx.doi.org/10.1109/ISPDC.2019.00010]

[19] O. Attia, I. Khoufi, A. Laouiti, and C. Adjih, "An Iot-blockchain architecture based on hyperledger framework for healthcare monitoring application", *2019 10th IFIP International Conference on New Technologies, Mobility and Security (NTMS),* pp. 1-5, 2019. [http://dx.doi.org/10.1109/NTMS.2019.8763849]

[20] R.H. Miller, and I. Sim, "Physicians' use of electronic medical records: Barriers and solutions", *Health Aff.,* vol. 23, no. 2, pp. 116-126, 2004. [http://dx.doi.org/10.1377/hlthaff.23.2.116] [PMID: 15046136]

[21] A.L. Terry, C.F. Thorpe, G. Giles, J.B. Brown, S.B. Harris, G.J. Reid, A. Thind, and M. Stewart, "Implementing electronic health records: Key factors in primary care", *Can. Fam. Physician,* vol. 54, no. 5, pp. 730-736, 2008. [PMID: 18474707]

[22] Y. Chen, S. Ding, Z. Xu, H. Zheng, and S. Yang, "Blockchain-based medical records secure storage and medical service framework", *J. Med. Syst.,* vol. 43, no. 1, p. 5, 2019. [http://dx.doi.org/10.1007/s10916-018-1121-4] [PMID: 30467604]

[23] R. Guo, H. Shi, Q. Zhao, and D. Zheng, "Secure attribute-based signature scheme with multiple authorities for blockchain in electronic health records systems", *IEEE Access,* vol. 6, pp. 11676-11686, 2018. [http://dx.doi.org/10.1109/ACCESS.2018.2801266]

[24] H. Wang, and Y. Song, "Secure cloud-based EHR system using attribute-based cryptosystem and blockchain", *J. Med. Syst.,* vol. 42, no. 8, p. 152, 2018. [http://dx.doi.org/10.1007/s10916-018-0994-6] [PMID: 29974270]

[25] A. Roehrs, C.A. da Costa, and R. da Rosa Righi, "OmniPHR: A distributed architecture model to integrate personal health records", *J. Biomed. Inform.,* vol. 71, pp. 70-81, 2017. [http://dx.doi.org/10.1016/j.jbi.2017.05.012] [PMID: 28545835]

[26] A.F. Hussein, N. ArunKumar, G. Ramirez-Gonzalez, E. Abdulhay, J.M.R.S. Tavares, and V.H.C. de Albuquerque, "A medical records managing and securing blockchain based system supported by a Genetic Algorithm and Discrete Wavelet Transform", *Cogn. Syst. Res.,* vol. 52, pp. 1-11, 2018. [http://dx.doi.org/10.1016/j.cogsys.2018.05.004]

[27] J. Yang, M. Onik, N.Y. Lee, M. Ahmed, and C.S. Kim, "Proof-of-familiarity: A privacy-preserved blockchain scheme for collaborative medical decision-making", *Appl. Sci.,* vol. 9, no. 7, p. 1370, 2019. [http://dx.doi.org/10.3390/app9071370]

[28] S. Khezr, M. Moniruzzaman, A. Yassine, and R. Benlamri, "Blockchain technology in healthcare: A comprehensive review and directions for future research", *Appl. Sci.,* vol. 9, no. 9, p. 1736, 2019. [http://dx.doi.org/10.3390/app9091736]

[29] X. Yue, H. Wang, D. Jin, M. Li, and W. Jiang, "Healthcare data gateways: found healthcare intelligence on blockchain with novel privacy risk control", *J. Med. Syst.,* vol. 40, no. 10, p. 218, 2016. [http://dx.doi.org/10.1007/s10916-016-0574-6] [PMID: 27565509]

[30] J.H. Frost, and M.P. Massagli, "Social uses of personal health information within PatientsLikeMe, an online patient community: what can happen when patients have access to one another's data", *J. Med.*

Internet Res., vol. 10, no. 3, p. e15, 2008.
[http://dx.doi.org/10.2196/jmir.1053] [PMID: 18504244]

[31] L. Castaldo, and V. Cinque, "Blockchain-based logging for the cross-border exchange of ehealth data in europe", *International ISCIS Security Workshop,* Springer: Cham, pp. 46-56, 2018.
[http://dx.doi.org/10.1007/978-3-319-95189-8_5]

[32] P.T. Duy, D.T.T. Hien, D.H. Hien, and V.H. Pham, "A survey on opportunities and challenges of Blockchain technology adoption for revolutionary innovation", *Proceedings of the Ninth International Symposium on Information and Communication Technology,* pp. 200-207, 2018.
[http://dx.doi.org/10.1145/3287921.3287978]

[33] X. Liang, J. Zhao, S. Shetty, J. Liu, and D. Li, "Integrating blockchain for data sharing and collaboration in mobile healthcare applications", *2017 IEEE 28th annual international symposium on personal, indoor, and mobile radio communications (PIMRC).*, pp. 1-7, 2017.
[http://dx.doi.org/10.1109/PIMRC.2017.8292361]

[34] R. Zhang, and L. Liu, "Security models and requirements for healthcare application clouds", *2010 IEEE 3rd International Conference on cloud Computing,* pp. 268-275, 2010.
[http://dx.doi.org/10.1109/CLOUD.2010.62]

[35] A.A. Omar, M.Z.A. Bhuiyan, A. Basu, S. Kiyomoto, and M.S. Rahman, "Privacy-friendly platform for healthcare data in cloud based on blockchain environment", *Future Gener. Comput. Syst.*, vol. 95, pp. 511-521, 2019.
[http://dx.doi.org/10.1016/j.future.2018.12.044]

[36] H. Kaur, M.A. Alam, R. Jameel, A.K. Mourya, and V. Chang, "A proposed solution and future direction for blockchain-based heterogeneous medicare data in cloud environment", *J. Med. Syst.*, vol. 42, no. 8, p. 156, 2018.
[http://dx.doi.org/10.1007/s10916-018-1007-5] [PMID: 29987560]

[37] X. Li, X. Huang, C. Li, R. Yu, and L. Shu, "EdgeCare: leveraging edge computing for collaborative data management in mobile healthcare systems", *IEEE Access,* vol. 7, pp. 22011-22025, 2019.
[http://dx.doi.org/10.1109/ACCESS.2019.2898265]

[38] H.T. Vo, A. Kundu, and M.K. Mohania, "Research Directions in Blockchain Data Management and Analytics", In: *In EDBT*, 2018, pp. 445-448.

[39] L. Zhu, Y. Wu, K. Gai, and K.K.R. Choo, "Controllable and trustworthy blockchain-based cloud data management", *Future Gener. Comput. Syst.*, vol. 91, pp. 527-535, 2019.
[http://dx.doi.org/10.1016/j.future.2018.09.019]

[40] P. Genestier, S. Zouarhi, P. Limeux, D. Excoffier, A. Prola, S. Sandon, and J. M. Temerson, "Blockchain for consent management in the ehealth environment: A nugget for privacy and security challenges", *J. Int. Soc. Telemed. eHeal.*, vol. 5, p. GKR-e24, 2017.

[41] A. Azaria, A. Ekblaw, T. Vieira, and A. Lippman, "Medrec: Using blockchain for medical data access and permission management", *2016 2nd International Conference on Open and Big Data (OBD).* (pp. 25-30). IEEE, 2016.

[42] C. Saraf, and S. Sabadra, "Blockchain platforms: A compendium", *2018 IEEE International Conference on Innovative Research and Development (ICIRD),* pp. 1-6, 2018.

[43] "Ethereum", https://www.ethereum.org/

[44] "Ripple", https://ripple.com/

[45] "Hyperledger", https://www.hyperledger.org/

[46] D.V. Dimitrov, "Blockchain applications for healthcare data management", *Healthc. Inform. Res.*, vol. 25, no. 1, pp. 51-56, 2019.
[http://dx.doi.org/10.4258/hir.2019.25.1.51] [PMID: 30788182]

[47] A. Panesar, *Machine Learning and AI for Healthcare.* Apress, 2019.

[http://dx.doi.org/10.1007/978-1-4842-3799-1]

[48] J.H. Frost, and M.P. Massagli, "Social uses of personal health information within PatientsLikeMe, an online patient community: what can happen when patients have access to one another's data", *J. Med. Internet Res.,* vol. 10, no. 3, p. e15, 2008. [http://dx.doi.org/10.2196/jmir.1053] [PMID: 18504244]

[49] L. Castaldo, and V. Cinque, "Blockchain-based logging for the cross-border exchange of ehealth data in europe", In: *International ISCIS Security Workshop* Springer: Cham, 2018, pp. 46-56. [http://dx.doi.org/10.1007/978-3-319-95189-8_5]

[50] P.T. Duy, D.T.T. Hien, D.H. Hien, and V.H. Pham, "A survey on opportunities and challenges of Blockchain technology adoption for revolutionary innovation", *Proceedings of the Ninth International Symposium on Information and Communication Technology,* pp. 200-207, 2018. [http://dx.doi.org/10.1145/3287921.3287978]

[51] X. Liang, J. Zhao, S. Shetty, J. Liu, and D. Li, "Integrating blockchain for data sharing and collaboration in mobile healthcare applications", *2017 IEEE 28th annual international symposium on personal, indoor, and mobile radio communications (PIMRC).* (pp. 1-5). IEEE, 2017. [http://dx.doi.org/10.1109/PIMRC.2017.8292361]

[52] V. Patel, "A framework for secure and decentralized sharing of medical imaging data *via* blockchain consensus", *Health Informatics J.,* vol. 25, no. 4, pp. 1398-1411, 2019. [http://dx.doi.org/10.1177/1460458218769699] [PMID: 29692204]

[53] A. Khatoon, "A blockchain-based smart contract system for healthcare management", *Electronics,* vol. 9, no. 1, p. 94, 2020. [http://dx.doi.org/10.3390/electronics9010094]

[54] S. Pirbhulal, W. Wu, and G. Li, "A biometric security model for wearable healthcare", *2018 IEEE International Conference on Data Mining Workshops (ICDMW),* pp. 136-143, 2018. [http://dx.doi.org/10.1109/ICDMW.2018.00026]

[55] S. Khezr, M. Moniruzzaman, A. Yassine, and R. Benlamri, "Blockchain technology in healthcare: A comprehensive review and directions for future research", *Appl. Sci.,* vol. 9, no. 9, p. 1736, 2019. [http://dx.doi.org/10.3390/app9091736]

CHAPTER 8

Design and Development of Blockchain Integrated IoT System for Pharmaceutical Applications

Jyothy S. T.[1,*] and **Mrinal Sarvagya**[1]

[1] *REVA University, Bangalore, India*

Abstract: The integration of the Internet of Things (IoT) technology in the pharmaceutical industry has the potential to bring about significant advancements and improvements, particularly in areas like supply chain management, drug discovery, and patient monitoring. By leveraging IoT technology in these ways, the pharmaceutical industry can enhance efficiency, improve product quality, ensure patient safety, and mitigate the risks associated with counterfeit drugs in the market. However, it was essential to address cybersecurity concerns to safeguard sensitive data and maintain the integrity of the IoT ecosystem in healthcare. The concept of integrating IoT with blockchain in the pharmaceutical supply chain is indeed a powerful solution to address the challenges faced by the industry, particularly in supply chain management and the prevention of counterfeit drugs. Integrating IoT with blockchain technology also results in many advantages like accurate location tracking, real-time updates and visibility, counterfeit prevention, timely delivery to customers, environmental parameter control, expiry date monitoring, pharmaceutical warehouse management, automated decision making and many more. By combining the strengths of IoT and blockchain, the pharmaceutical industry can build a robust and secure supply chain ecosystem. This integrated approach not only enhances transparency and traceability but also addresses critical issues such as counterfeiting by ensuring the delivery of safe and authentic pharmaceutical products to consumers. Environmental parameters are also monitored in a warehouse to avoid spoilage of medicines. The date of expiry of medicines and related equipment are monitored and categorized into three levels such as discard the medicines if expired, near to expiry if the expiry date is nearer and store in the warehouse if the date of expiry is far, and a message is sent accordingly to the supervisor for necessary action.

Keywords: Blockchain, Internet of things (IoT), Pharmaceutical industries, Radio-frequency identification, Supply chain management.

* **Corresponding author Jyothy S. T.:** REVA University, Bangalore, India; E-mail: jyothyst@reva.edu.in

Santosh Kumar Das, Soumi Majumder and Nilanjan Dey (Eds.)

INTRODUCTION

Internet of Things (IoT)

Pharmaceutical drugs have become an integral part of life nowadays in diagnosing and treating patients. In recent years, these essential pharmaceutical drugs and other medical operators and devices have become more susceptible to pharmaceutical crime. Counterfeiting of drugs *i.e.*, replacement of standard drugs and vaccines with substandard products, has increased nowadays, which risks the life of patients [1, 2].

Counterfeiting of drugs have a negative impact on consumers – diseases may not be cured, or future diseases may not be prevented. This results in a loss of confidence in healthcare sectors and service providers, resulting in the expenditure of more money in the future as the disease is not cured completely. There are chances of a shortage of income due to illness. Counterfeiting of drugs has a more negative impact on producers, like reduction in sales, amount and efforts made in branding activities, loss of reputation gained so far, the probable efforts in terms of both cost and manpower required for the disposal of counterfeit pharmaceutical drugs and medical products, litigation costs involving these drugs and products and majorly on people who were unknowingly victimized by counterfeits. Counterfeit pharmaceutical drugs and medical products redirect the resources away from actual treatment, stealing from limited health budgets that are already short of resources [3]. At the same time, counterfeit pharmaceutical drugs and medical products can affect losses in corporate taxes and VAT, increase managerial and enforcement costs to protect the supply chain from theft of medicines carried in vehicles from warehouses to consumers and increase healthcare costs to treat the adverse effects of counterfeit pharmaceutical drugs and medical products.

Pharmaceutical warehouses need exceptional care in maintaining drugs and other products safely as these products are temperature sensitive and also required to be maintained under specific conditions for certain environmental parameters such as temperature, humidity, pressure, light, oxygen, *etc* [4, 5]. If the drugs are not maintained under specified conditions, drugs will spoil and cannot be used for any purpose. This will incur a financial loss to both producers *i.e.*, pharmaceutical industries and the government. This will result in improper usage of resources. To avoid this, there should be some mechanism designed to monitor and control the environmental conditions regularly as required for a particular pharmaceutical drug or any other medical device related to healthcare. If there is any variation in the values sensed by IoT monitoring sensors from the specified values, then the designed system should automatically control or report to the supervisor for necessary action.

The designed system should also monitor the expiry date of pharmaceutical drugs and medical equipment. Because of the expiry of pharmaceutical drugs and medical equipment, a lot of medicines and other products are wasted. This, in turn, causes a loss of revenue to the pharmaceutical industries and also to the nation. The designed system should continuously monitor the expiry date of drugs and medical-related devices and classify them into three categories based on the expiry date. Three categories are expired, near to expiry, and store in the warehouse.

If a particular drug or medical-related device falls under the expired category, it indicates that the drug or device has already expired and needs to be discarded as usage of these products will risk the lives of humans. If any drug or medical device falls under the near to expiry category, they are immediately circulated for distribution. This avoids spoilage of drugs and prevents the loss of revenue to the pharmaceutical manufacturing company and the government. If the expiry date is more than one year, then the particular product or drug can be stored in the warehouse.

For all the above-mentioned categories, an alert is issued by a system to the supervisor for necessary action. Manual verification of the expiry of drugs and other medical-related devices is very difficult and requires a lot of manpower. Hence, an automated verification system to detect the expiry date of drugs and other medical-related devices is required. This avoids spoilage of medicines and proper use of medicines and drugs without affecting any lives and loss of revenue to the pharmaceutical companies and the nation.

A system needs to be designed for industrial mechanics and maintenance and to store the drugs safely by controlling the environmental conditions. In case any machinery breaks down in a large warehouse, an alert is made by a system that monitors environmental parameters continuously. Manual monitoring of machinery is very difficult and requires manpower to complete the task. In the manual monitoring system, the supervisor will not be able to know immediately about machine failure. This will cause a delay in identifying the machinery failure and repairing the machinery. Until the machinery is serviced, production is halted. This, in turn, causes delays in the supply of urgent drugs and other medical products, resulting in a loss of revenue to both pharmaceutical industries and the government.

To make the supply chain transparent and automate the verification of drugs and other medical-related devices and monitoring of machinery, there is a requirement to transform the current pharmaceutical industry [4, 6]. The Internet of Things has the power to revolutionize the current pharmaceutical supply chain and warehouse management in the process ranging from drug discovery to remote patient access and monitoring [7]. Using IoT sensors, continuous monitoring of environmental parameters can be sensed, and a system needs to be developed to take necessary action if there are any variations in the sensed environmental parameters. IoT sensors are also used to verify the expiry date of the drugs, classify them into the three categories and take the necessary actions like discarding medicines, immediate circulation and store in the warehouse. IoT sensors are used to monitor and store the working condition of the machinery, and an alert is made to the concerned authority for necessary action to avoid delay in production in case of machinery failure. IoT sensors are used to monitor environmental parameters throughout the supply chain, right from the warehouse to the consumers. During the transportation of drugs and other medical-related products, environmental parameters such as temperature, humidity, light, pressure and oxygen [8] are monitored and controlled if there are any variations, as during transportation, medical drugs and products experience different environmental conditions and need to be taken care of to avoid spoilage of drugs and other medical-related products.

A huge amount of data is generated when the entire process is connected to IoT. A lot of sensors are required in the entire process of the supply chain, from the warehouse to consumers, and warehouse management to sense environmental parameters and enable the entire process as a smart system. These sensors generate a huge volume of data. The generated data is required by devices for decision making, to carry out the next process or to report to the supervisor for necessary action. When many devices and sensors are connected to IoT and a huge amount of data is generated, few devices require data generated by few sensors. In this case, there are chances of data breaching – data is stolen without the knowledge of an authorized person. In pharmaceutical applications, providing security is the most important as we need to protect the personal data and health records of all the patients. Privacy and security of all data generated by the sensors and devices in the entire process are provided by the implementation of blockchain.

Blockchain

Blockchain technology acts as a distributed ledger technology where the transactions are made in digital mode, and these digital mode transactions are recorded, verified, and validated throughout the network of nodes, such as without the acceptance of central authority [9 - 11]. Distributed ledgers provide transparency, trust, and data security [12, 13].

Blockchain is categorized into three types. They are Public, Private and Consortium [14]. A public blockchain type is like a bitcoin in which anyone in the world who is a part of the blockchain has the right to access data [15]. A private blockchain is one where a single person in the entire blockchain has the absolute right to verify the transaction as well as add a new block to the blockchain. In a consortium blockchain, in place of one person, there is a group of people who can verify the transactions and add these transactions to the blockchain.

Pharmaceutical applications cannot be implemented using a public blockchain because, in the public blockchain, those who are part of a particular blockchain are allowed to access the data. Privacy and security cannot be provided in this type of implementation. However, providing privacy and security is most important for pharmaceutical applications. A private blockchain provides privacy and security for the data as a single person has the complete right to access the data and to complete the verification process and the task of adding new transactions into the system. However, pharmaceutical industries are very big, and a huge volume of data is generated in the entire process. A single person managing the process is a tedious job, and implementation is impossible. Consortium blockchain can be implemented for pharmaceutical applications, where rather than one person, there will be a cluster of people who manage the verification of transactions and add new transactions into the system.

Components of a Blockchain Architecture

The key components of blockchain architecture consist of Node, Transaction, Block, Chain, Miners, and Consensus. Node is a computer in the blockchain architecture. Transaction is a data record verified by the authorized nodes or persons in the blockchain. With respect to pharmaceutical applications, transactions include data generation by sensors, monitoring of vital parameters such as BP, sugar, ECG, *etc.* of the patient, purchasing requisition of pharmaceutical drugs or medical-related devices by consumers, delivery of ordered items such as drugs and components related to healthcare by pharmaceutical industries, prescription given by doctors to patients or financial transaction such as payment of fee by patients or purchase of drugs or salary component, *etc.*

Block is a sealed data box comprising of:

- A native hash code of the respective block, which is calculated based on the value available in the current block.
- Hash code from the previous block in the sequence of blocks to link these blocks in a blockchain network.

- A set of timestamped transactions that records the time at which the transaction takes place and other information related to the particular transaction.

A chain is an ordered sequence of blocks in a blockchain network to link the blocks in a blockchain network.

Miners are the nodes that verify and validate each and every transaction before they are added to the blockchain network. Not all the nodes act as miners. Certain nodes are identified to act as miners in a blockchain structure.

Consensus comprises certain rules and regulations like agreement for performing blockchain operations like verifying and validating new transactions, adding new transactions, adding new devices, removing devices, identification of miners, linking blocks, *etc.* All the transactions should abide by the rules and agreements set by the particular blockchain network.

Each block of a blockchain comprises data of the block, the hash of the block, and the hash from the previous block to link the blocks in a blockchain network. The structure of a block is as shown in Fig. (**1**).

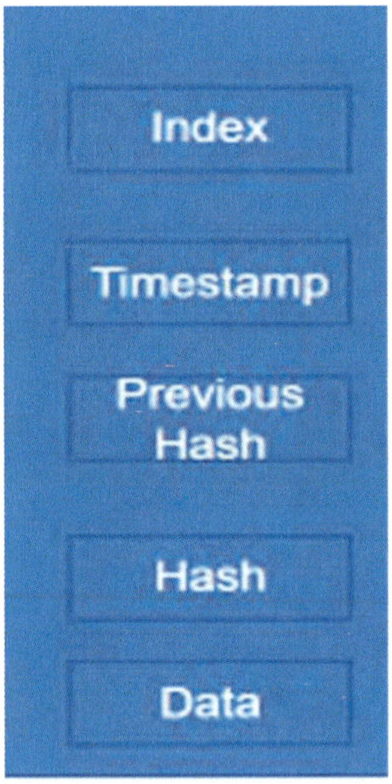

Fig. (1). Structure of an individual block in a blockchain.

The chain of blocks is shown in Fig. (**2**).

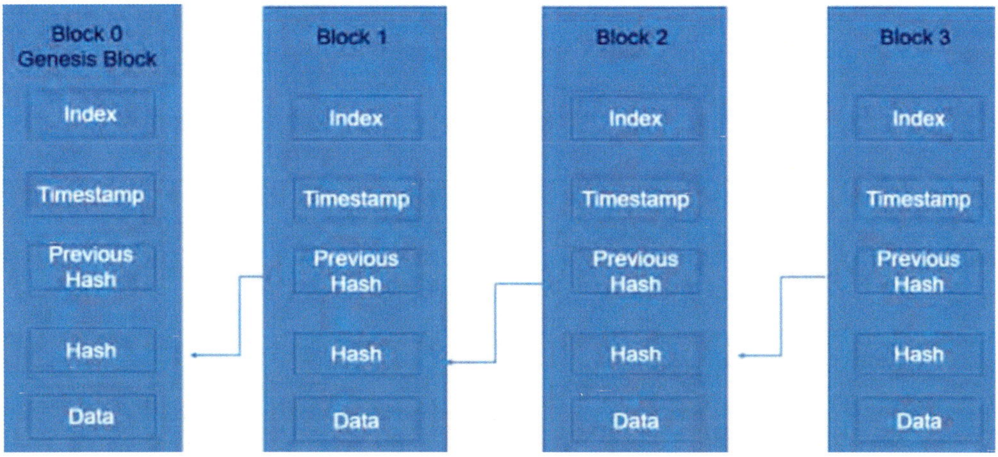

Fig. (2). Chain of blocks.

In the chain of blocks shown in Fig. (**2**), the first block *i.e.*, block 0, is called the genesis block, whose previous hash value is set to zero [14].

LITERATURE SURVEY

Alemuda *et al.* proposed a system that integrated one M2M, IoT, and big data analytics [1]. This proposed system reflects a contemporary and comprehensive approach to industrial automation, particularly in the pharmaceutical sector. The emphasis on real-time monitoring, data-driven decision-making, and preventive maintenance aligns with the industry's best practices for efficiency and reliability. The information provided offers a glimpse into how modern technologies are applied to address challenges in manufacturing environments. In summary, the proposed system exhibits a forward-thinking approach by utilizing the one M2M standard, emphasizing automation, data standardization, and proactive maintenance in the pharmaceutical sector. The incorporation of big data analytics reflects an understanding of the challenges and opportunities associated with managing and deriving value from large volumes of data in an industrial setting (2019).

Yousuf and Svetinovic described that in spite of notable research, there are still challenges in the supply chain management system that need to be resolved [2]. The challenges that need to be addressed are information about the product transported, the transport history of the product, *etc.* In this paper, the authors focus on identifying the issues that occur frequently and their conditions or requirements for the implementation of integrating blockchain. A conceptual overview is created by the authors to encourage young researchers to carry out research and development in the field of blockchain (2019).

Wazid *et al.* proposed a recently developed authentication policy for medicine or drug anti-counterfeiting design in the IoT domain [3]. The suggested scheme is secure against different known attacks, such as counterfeiting or distribution of fake and substandard medicines. The formal security authentication system with the powerful and universally accepted tool *i.e.*, the AVIPSA tool, shows that the said design provides security. Their design is cost-effective and comparable in the aspects of computational speed and cost incurred during communication and provides additional aspects as compared to other currently available methods. They have experimented with the suggested method using the tool NS_2 simulator, and the results of the simulation exhibit the attainability of the design. Finally, a good tradeoff between securities and other extra functionality characteristic features and efficiency indicates that the suggested method is appropriate for the eradication of counterfeiting of pharmaceutical drugs, medicines, and vaccines. In summary, the described authentication policy by Wazid *et al.* appears to be a well-rounded solution, addressing security concerns related to counterfeit medicines in the IoT domain. The use of formal security assessment tools, simulation results, and comparisons with existing methods contribute to the credibility of the proposed design (2017).

Alagarsamy *et al.* highlight that pharmaceutical companies have begun utilizing IoT in their manufacturing processes, and the potential for IoT in the healthcare sector is emphasized globally [4]. The implementation of IoT technology is expected to bring benefits to pharmaceutical companies, especially in terms of low-energy and low-cost IoT networks. The author also mentions that low-energy, low-cost IoT networks and solutions are still in the early stages of development. The significance of IoT is outlined in various areas, including pharmaceutical manufacturing, drug discovery processes, and clinical trials (2019).

Dey N explains the data-driven approach for bio-medical and healthcare applications and their challenges and also the implemention of the Internet of Medical Things and translational bioinformatics to store personal health records and electronic health records by enabling privacy and security techniques [5].

Rejeb *et al.* summarize a research study that reviews the capability of using the Internet of Things (IoT) in managing supply chain systems (SCM) [6]. The author concentrates on the major issues such as bibliometric analysis, results of the analysis, areas of exploration, and significance of researchers. In summary, the research study appears to contribute to the growing body of knowledge on the use of IoT in supply chain management and logistics. It not only highlights the significance of IoT in these domains but also aims to guide future research efforts by identifying areas where further exploration and innovation are needed (2020).

Kelly *et al.* summarize the capability of IoT in the healthcare sector as an emerging trend that holds significant potential for advancing medical practices and improving patient care [7]. Ongoing research in this field offers a tremendous opportunity for healthcare providers to predict and analyze health-related issues, enhance treatment methodologies, and enable remote monitoring of patients, including tracking specific parameters such as blood pressure and heart rate. The capabilities in the healthcare sector include prediction and analysis, remote monitoring, flexible supply chain management, technology-assisted healthcare systems, safety, privacy and security of the patients and healthcare systems, data management and cybersecurity, and filling research gaps. IoT in healthcare offers transformative possibilities, ranging from improved patient monitoring to enhanced supply chain management. While acknowledging the potential benefits, researchers continue to explore and address challenges to ensure the responsible and effective integration of IoT technologies in the healthcare landscape.

Mohsin and Yellampalli highlight the significance of monitoring and controlling temperature in the pharmaceutical cold chain, encompassing the storage and transportation of drugs from suppliers or distributors to consumers [8]. The paper highlights the major points on the preservation of the quality of cold storage products like vaccines and medicines. The authors discuss the technical aspects, packing materials, and security concerns. Also, the authors discuss the coordination challenges between the warehouse, supply chain logistics and customers. The authors summarize the research and underscore the critical role of proper temperature control and management of the pharmaceutical cold chain. The study draws attention to existing coordination challenges and emphasizes the importance of improving cold chain management practices to ensure the safety, security, and efficacy of temperature-sensitive pharmaceuticals (2017).

Aich *et al.* highlight the benefits of using IoT integrated blockchain-based supply chain management over conventional supply chain management systems, which we have been using till now [9]. This work outlines the difficulties faced by some of the industries like automotive, pharmaceutical, food, and retail manufacturers by implementing the traditional supply chain system and the solutions to the issues provided by the blockchain and IoT-based systems in these domains. Two emerging technologies, IoT and blockchain, in the pharmaceutical supply chain management system, can eradicate the problems faced by various industries and make the design more efficient, secure and trustworthy. Ultimately, as an example, a case study of the seafood industry supply chain system has been proposed to make use of the benefits of integrating blockchain into the network. The proposed system will assist the manufacturing sector individuals to analyze the advantages of the blockchain integrated IoT-based supply chain system and to make an attempt to experiment with those in real life to get better results.

Premkumar and Srimathi highlight the benefits of adopting an integrated supply chain management system that combines the Internet of Things (IoT) with blockchain technology, in contrast to traditional supply chain management systems [10]. They also highlight the advantages of moving from traditional supply chain management systems to an integrated approach that leverages both IoT and blockchain technologies. The focus on the pharmaceutical supply chain and the seafood industry case study demonstrates the potential for improved efficiency, security, and trustworthiness in supply chain operations across various sectors. The proposal encourages real-world experimentation to validate the benefits of the proposed system (2020).

Bhalerao *et al.* explain the concept of blockchain used in the transport of dairy products such as milk, curd, ghee, sweets, *etc* [11]. All the physical and environmental parameters like sender, position, location, temperature, humidity and many more are recorded at every instant. The end user gets complete information about the product from packing until it reaches the user. The authors highlight the eight major characteristics to be included in developing Hyperledger Fabric. They are Blockchain Architect, Blockchain User, Blockchain Regulator, Blockchain Developer, Blockchain Operator, Membership Services, Traditional Processing Platform and Traditional Data Coordinator (2019).

Sun W describes the scope and development of blockchain in recent years that has given rise to a lot of innovations in the field of supply chain and digital finance [12]. Also, the author describes the use of smart contract technology in double-chain finance systems, which confirms the transaction and instruments. The technology presented in this paper motivates the researchers and developers in this area to use the concept in their work and implementation (2022). Muessigmann *et al.* presented surveys of different papers on blockchain technology and supply chain research [13]. They addressed the employment of blockchain in logistics and supply chain management systems. The bibliometric analysis method is used in this paper. From co-citation analysis, this paper categorizes the present literature into five research groups, including theoretical sensemaking, conceptualizing and testing applications of blockchain, framing blockchain technology into supply chains, the technical outline of blockchain technology applications for real-world logistics and supply chain management uses, and a major part of blockchain in digital supply chain systems (2020).

Ratta *et al.* highlight the broad applicability of the Internet of Things (IoT) across diverse sectors, including agriculture, chemical, construction, healthcare, finance services, and smart cities [14]. In the healthcare sector, IoT finds applications in monitoring and controlling patients' health conditions regularly and ensuring the traceability of drugs. However, the review acknowledges that IoT comes with safety, privacy, and security concerns, which are addressed through the integration of advanced technologies, like IoT and Blockchain. In summary, the review recognizes the broad applicability of IoT and the specific challenges it poses in terms of safety, privacy, and security. The integration of Blockchain is proposed as a solution, particularly in the healthcare sector, to address these concerns and potentially revolutionize pharmaceutical supply chain management. The combined use of IoT and Blockchain holds significant promise for advancing processes in the healthcare industry (2021).

Das *et al.* propose the method peer-to-peer communication system. Various techniques like multi-objective optimization, game theory technique, non-linear geometric programming, and intuitionistic fuzzy logic are used in the proposed technique [15]. This model helps to achieve intelligent communication between source nodes and destination nodes.

Lee *et al.* outline a research effort focused on the application of the Internet of Things (IoT) technology in the healthcare sector, specifically in healthcare asset management [16]. The authors stress the decade-long technological analysis of healthcare data, including patients' and pharmaceutical data, technological components, systems for assessment, improvements required, scalability and future applications. The research highlights the advancements in healthcare asset management through the integration of IoT technologies. The proposed system, involving ANN, RFID, WSN, and FL components, demonstrates improved efficiency in terms of manpower, patient safety, and maintenance planning. The scalability and potential for future applications suggest that this model could be extended to benefit a broader category of products or hospitals in the ongoing evolution of IoT technologies in healthcare (2015).

Dass A. K. and Das S. K. explain the capability of wireless sensor networks (WSN) and their applications in various sectors [17]. The authors also discuss the process of cryptography and how artificial neural networks can increase the security and throughput of the network.

Saindane *et al.* indicate that pharmaceutical industries are considering blockchain as an effective solution for tracking and tracing pharmaceutical drugs and products throughout the supply chain [18]. The authors describe the benefits of counterfeit prevention, reduced overhead costs, and enhanced security and privacy for IoT. In summary, the pharmaceutical industry is exploring blockchain as a comprehensive solution for improving transparency, security, and efficiency in the supply chain. The proposed design focuses on leveraging blockchain's features to enhance tracking and tracing applications, ensuring the authenticity of pharmaceutical products and contributing to the overall security and privacy of IoT components within the supply chain (2020).

Raj and Sowmiya describe the benefits provided by blockchain technology to the user along with a few concerns, such as data breaching, decentralization, privacy and security [19]. In blockchain, each and every data is timestamped means information like data, transaction details, time and other information related to the application are recorded and saved within the system. There are some algorithms that enable the system to follow the rules and regulations framed by the particular blockchain. Blockchain technology is used mainly in e–commerce and transportation, which is completely automated. Since there are chances of data breaching, blockchain is used to provide security for the automated system (2021).

Sharma and Singh highlight the significance of the Internet of Things (IoT) in the pharmaceutical and healthcare sectors, particularly in manufacturing, warehousing, distribution, and supply chain management [20]. The authors summarize the current and future role of IoT in the pharmaceutical and healthcare sectors. They emphasize the ongoing improvements in IoT technology and anticipate its potential to bring about significant positive changes, including enhanced safety, reliability, and cost-effectiveness in the production and distribution of pharmaceutical products. The diverse applications of IoT in areas such as big data management, temperature monitoring, and counterfeit prevention showcase its multifaceted contributions to the pharmaceutical industry.

Bocek *et al.* indicate that numerous new companies are considering the adoption of blockchain-based solutions to establish cost-effective models and streamline operations by reducing bureaucracy [21]. The success of these companies in implementing IoT-integrated blockchain systems across various sectors will determine the practical utilization of the technically achievable advantages and characteristics. In summary, the focus is on the emergence of companies opting for blockchain-based solutions with the aim of cost-effectiveness and reduced bureaucracy. The success rate of IoT-integrated blockchain systems in real-world applications will shed light on the practical implementation and utilization of the advantages offered by these technologies across various sectors (2017).

Sudha *et al.* discuss the requirement for transformation in the existing supply chain used for agricultural products [22]. The tracking mechanism is not present in the existing supply chain system. No one can get information about products and their quality once they are transported. There are chances of spoilage of the products if they do not reach the consumers on time. This will cause financial loss both to the producers and the nation. The efforts made by producers will go to waste. This leads to the requirement for a tracking mechanism that enables both producers and consumers to track the items transported. The authors discussed the new technology that can be used for tracking mechanisms *i.e.*, the blockchain technology, which obtains transparency about the status of items transported and also improves the trust between the sender and receiver (2021).

Agarwal *et al.* describe the major role played by supply chain in companies and industries [23]. There are some issues in traditional supply chain management, such as privacy and security of the devices and data, transparency, traceability, delay in delivery of goods, counterfeiting of goods, *etc.* To overcome the shortcomings of traditional supply chain management, authors have come up with the blockchain as a feasible solution. They analyzed the different applications where blockchain can be used to provide privacy and security for all transactions. Issues and research challenges are described to motivate new researchers to take up this topic as their research area and contribute to society (2022).

Abdelgalil *et al.* discussed that the use of blockchain in industries is gaining importance nowadays [24]. In this paper, security aspects, strengths, weaknesses, advantages and vulnerabilities are discussed. A case study of a frozen food factory is provided, which gives the details of the idea for the development of the system and the constraints of the requirements. The authors also highlight that this case study can be taken as a reference for any other applications (2021). Sathya *et al.* discussed the scope of blockchain technology and its implementation in different sectors such as healthcare, banking applications, food sector, education and many more [25]. In this paper, blockchain technology is implemented in the food sector to connect farmers, buyers and sellers. Blockchain technology is implemented and experimented with in this paper with a detailed discussion on decentralization blockchain, security, privacy, tamperproofing and immutability for the supply chain management. The work proposed here uses smart contracts on Ethereum so that falsification of data and corruption can be reduced (2021).

Botcha *et al.* designed a system to increase visibility in the healthcare pharmaceutical supply chain using Blockchain-integrated Internet of Things (IoT) devices, which involves creating a secure, transparent, and traceable ecosystem [26]. Implementing this system would contribute to a more secure and transparent pharmaceutical supply chain, addressing the identified research gaps and enhancing overall efficiency, trust, and consumer protection in the industry –(2019 Top of Form).

Hader *et al.* highlight that the scope of blockchain technology extends from finance to other applications like the retail market [27]. Retail market design requests will always keep changing as per customer needs and to gain customer confidence. As per the changes, retailers request blockchain technology expertise to design to improve their performance through the supply chain system. They also analyze the impact of using blockchain technology in the retail market and obtain possible solutions to implement this emerging technology to increase customer loyalty. The implementation of blockchain technology in the retail sector needs multidisciplinary studies and the involvement of retailers to frame rules and regulations on the implementation of blockchain. The work carried out in this paper has given direction to young researchers in this area (2020).

Darbari S describe show the widespread use of technologies like the Internet and wireless sensor networks has given way to the extensive use of the applications of the Internet of Things [28]. The devices used in IoT structures have become an integral part of every individual's daily life. Automated services serviced by these smart IoT devices are purchased by a particular vendor or service provider. The service provider has the right to access any kind of data of each and every individual. There comes the importance of integrating blockchain with the Internet of Things. Blockchain applications provide a lot of benefits, such as decentralization, transparency, trustworthiness and immutability. This paper describes the concept of integrating two emerging technologies, IoT and blockchain, where privacy and data security are major concerns (2021).

Majumder and Dey discuss the method of knowledge management, which covers the creation, distribution, and maintenance of knowledge [29]. The work focuses on AI-empowered knowledge management to improve processes and acts as a reference resource for researchers and practitioners in academia and industry.

Das *et al.* present an intrusion detection system (IDS), which is capable of monitoring network traffic for suspicious activities and issues when such activities are located [30].

Das S K presents the applications of smart devices and technologies and their requirement and purposes. The usage of smart devices and their challenges are discussed [31].

Jyothy S T *et al.* outlined that an approach to leveraging the Internet of Things (IoT) for supply chain management in the pharmaceutical industry is well-considered and addressed critical challenges for standardizing supply chain management in the pharmaceutical industry [32]. Finally, highlighting the regulatory aspects and standards related to pharmaceutical supply chain management and IoT can add depth to the analysis.

IOT ENABLED TECHNOLOGIES

Following technologies enable the use of the Internet of Things.

Wireless Sensor Networks (WSN)

Wireless Sensor Networks (WSNs) consist of distributed devices integrated with sensors [16, 17]. WSNs are used to monitor the environmental and physical conditions. This network comprises a number of end nodes and routers and a coordinator. Numerous sensors are attached to end nodes. Data packets are routed from end nodes to the coordinator by end nodes or routers. The coordinator accumulates the data from all the nodes and connects the WSN of the internet.

Sensors

Sensors are used to monitor and control environmental and physical parameters to store drugs and medical equipment safely in a warehouse and also throughout the supply chain and retailers, until the drugs reach the patients.

Actuators

An actuator is a mechanical element of a machine that manages the movement or control of a mechanism or system. A control signal for an actuator is provided by a processor based on the inputs generated by sensors attached to end nodes.

Cloud Computing

Cloud computing enables us to store huge volumes of data generated by many devices and sensors over a network.

Big Data Analysis

As many devices are connected to the network, a huge volume of data is generated. Storage, management, processing and analysis of data is a tedious process using traditional databases and data processing tools. This led to the implementation of Big Data Analysis.

Communication Protocols

Communication protocols frame rules and regulations to be followed for the exchange of data among devices over a network. Communication protocols are the key component of an IoT system, which enable connectivity over a network. IoT protocols supported in each layer are listed in Table **1**.

Table 1. List of protocols.

Layer	Protocols
Application Layer	Hypertext Transfer Protocol HTTP, Constraint Application Protocol, WebSocket, Message Queuing Telemetry Transport, Extensible Messaging and Presence Protocol, Data Distribution Service, Advanced Message Queuing Protocol.
Transport Layer	Transmission control protocol (TCP) drives reliable data transfers, while user datagram protocol (UDP) prioritizes speed and efficiency.
Network Layer	Internet Protocol Version 4, Internet Protocol Version 6, IPv6 over Low-Power Wireless Personal Area Networks.
Link Layer.	Ethernet, WiMAX, WiFi, LR-WPAN, 2G/3G/ LTE - Cellular.

Embedded Systems

Embedded systems are designed to perform a specific task consisting of both hardware and software. Key components of an embedded system include processor, memory, networking units, input/output units and storage. Key component processors include microprocessors, microcontrollers, digital signal processors, System on Chip, *etc.* Depending on the complexity of applications, these processors are selected. For simple applications, microprocessors are used, and for complex applications, digital signal processors or System on Chip are used. Memory is an integral part of any embedded system used to store programs and variables. Programs are stored in permanent memory such as PROM, EPROM, EEPROM, Flash, NV- RAM, *etc.* Variables during execution are saved in temporary memory like Dynamic RAM or Static RAM, which are erased once the power is lost.

Radio-Frequency Identification

Radio-Frequency Identification (RFID) consists of an RFID reader and RFID tags [16]. RFID reader verifies whether the RFID tag is valid or not. RFID tags are used to track and trace the products right from the warehouse.

CHALLENGES FACED BY TRADITIONAL SUPPLY CHAIN

The block diagram of a traditional or conventional supply chain system is shown in Fig. (**3**).

In traditional supply chain systems shown in Fig. (**3**), there are no tracking mechanisms [18]. During the transport of medical products from warehouse to customers, there are chances of counterfeiting of drugs. This is the major drawback of the traditional supply chain system. A few of the major challenges faced by traditional supply chain systems are listed below [4, 9, 19]:

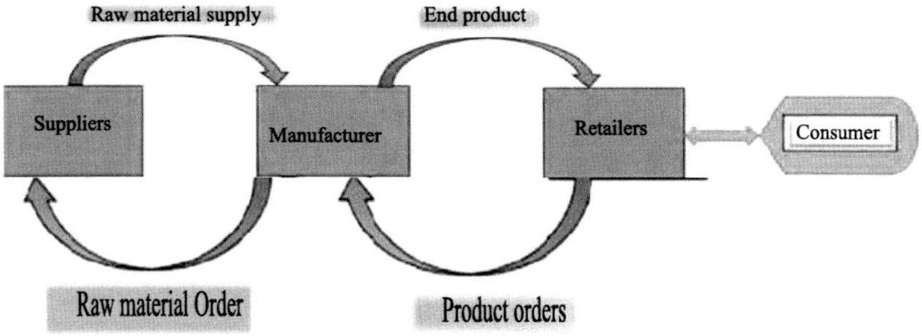

Fig. (3). Traditional supply chain system.

Transparent Supply Chain

There is no monitoring system in the entire supply chain right from the warehouse to retailers. There are chances of theft of medicines or counterfeiting of drugs [20, 7].

Temperature Control

Almost all medical products are temperature-sensitive [21]. During the transport of medical products through air, water, rail or road, they are exposed to different variations in the temperature and other environmental parameters. Due to variations in the temperature in the specified range, there are chances of spoilage of medicines and drugs.

Warehouse Management

A huge quantity of pharmaceutical drugs and other medical-related components are stored in warehouses. Manual monitoring of environmental and physical parameters like temperature, light, humidity and light is difficult as the warehouse is not automated in the traditional warehouse monitoring system [20].

Lack of Standardization

There are no standard communication protocols framed in traditional systems to exchange data between devices and nodes in a network.

SOLUTIONS PROVIDED BY BLOCKCHAIN-INTEGRATED INTERNET OF THINGS SUPPLY CHAIN SYSTEM

Blockchain-integrated IoT supply chain management system is shown in Fig. (**4**) [9, 22, 23]. In this structure, all the transactions are recorded from the placing of the order by individuals/pharmacies/hospitals to pharmaceutical industries to the delivery of ordered pharmaceutical drugs and medical products to consumers. This tracking and tracing will avoid the counterfeiting of drugs. As the entire system is automated by sensing environmental and physical parameters, the spoilage of drugs, medicines and vaccines is avoided by the automatic control of variations in environmental and physical parameters than that prescribed [24, 25].

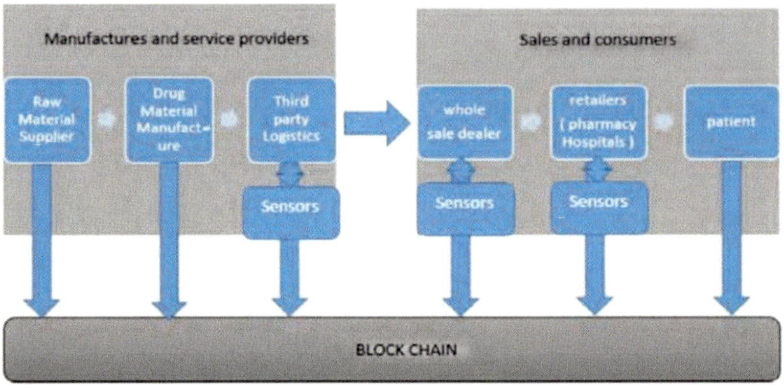

Fig. (4). Blockchain-integrated IoT supply chain system.

Solutions provided by blockchain-integrated Internet of Things supply chain systems are listed below.

Transparent Supply Chain

Transparent Supply Chain is obtained by IoT sensors and Blockchain, which records each and every transaction [7, 26, 27]. RFID sensors and GPS are used to track and trace the medicines and medical equipment throughout the supply chain [16, 28]. The implementation is shown in the flowchart shown in Fig. (**5**). Whether the container lock is opened and locked throughout the journey is determined by the LDR sensor.

Temperature Control

Most of the pharmaceutical products are temperature-sensitive. Temperature sensors monitor and control temperature continuously as required by the products [29]. The system can be designed to monitor and control other environmental and physical parameters like humidity, oxygen, pressure, *etc* [21]. The implementation of monitoring of environmental parameters is shown in Fig. (**6**). If there are any variations in the sensed parameters, a supervisor is alerted to take necessary action.

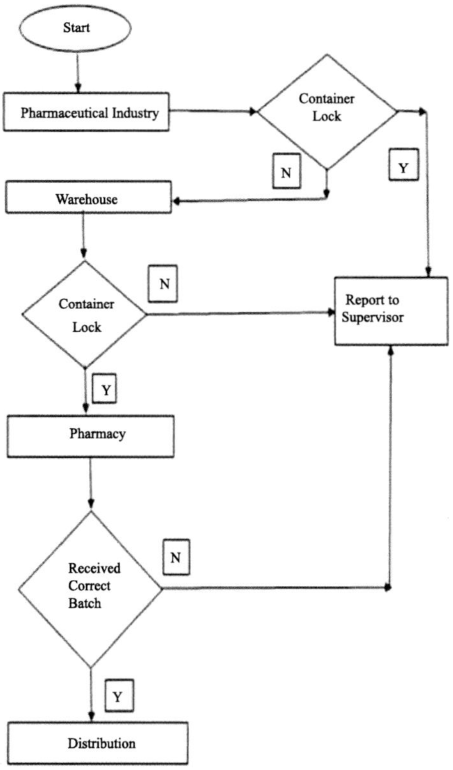

Fig. (5). Flowchart to ensure proper delivery of smart pharmaceutical courier box without counterfeit.

Smart Warehouse

IoT sensors monitor and control environmental and physical parameters as required by the medicines and medical equipment [30 - 32].

Standardization

A proper code of conduct framework can be made for all pharmaceutical industries.

Telemedicine

Remote patient monitoring, diagnosis and treating patients can be implemented [7].

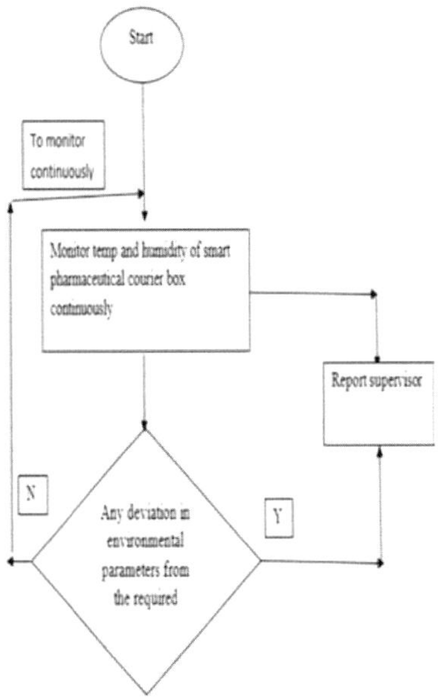

Fig. (6). Flowchart to monitor environmental parameters.

CONCLUSION

This chapter provides a detailed analysis of the use of IoT and Blockchain technologies in pharmaceutical applications. IoT sensors are used to monitor various environmental and physical parameters. This concept is used to track and trace the position of each medicine and store the medicines in the warehouse and supply chain safely without spoilage. Blockchain technology is implemented to provide transparency for all the data available in the system. Among private, public and consortium kinds of blockchain, consortium blockchain is used for the pharmaceutical sector. In consortium blockchain, instead of one person, there will be group of people managing all the transaction validation and verification processes. This chapter also gives a comparison of traditional and blockchain-integrated IoT supply chains and the advantages of using blockchain-integrated IoT supply chains.

REFERENCES

[1] F. Alemuda, I. Alinursafa, D. Adrianto, and I.K.A. Enriko, "Incorporating one M2M platform to pharmaceutical industry", *2019 IEEE International Conference on Internet of Things and Intelligence System (IoTaIS),* IEEE., pp. 36-41, 2019. [http://dx.doi.org/10.1109/IoTaIS47347.2019.8980431]

[2] S. Yousuf, and D. Svetinovic, "Blockchain technology in supply chain management: preliminary study", *2019 Sixth International Conference on Internet of Things: Systems, Management and Security (IOTSMS),* IEEE., pp. 537-538, 2019. [http://dx.doi.org/10.1109/IOTSMS48152.2019.8939222]

[3] M. Wazid, A.K. Das, M.K. Khan, A.A.D. Al-Ghaiheb, N. Kumar, and A.V. Vasilakos, "Secure authentication scheme for medicine anti-counterfeiting system in IoT environment", *IEEE Internet Things J.,* vol. 4, no. 5, pp. 1634-1646, 2017. [http://dx.doi.org/10.1109/JIOT.2017.2706752]

[4] S. Alagarsamy, R. Kandasamy, L. Subbiah, and S. Palanisamy, "Applications of internet of things in pharmaceutical industry", *SSRN,* p. 3441099, 2019.

[5] N. Dey, Ed., *Data-Driven Approach for Bio-medical and Healthcare.* Springer Nature, 2022.

[6] A. Rejeb, S. Simske, K. Rejeb, H. Treiblmaier, and S. Zailani, "Internet of things research in supply chain management and logistics: A bibliometric analysis", *Internet of Things,* vol. 12, p. 100318, 2020. [http://dx.doi.org/10.1016/j.iot.2020.100318]

[7] J.T. Kelly, K.L. Campbell, E. Gong, and P. Scuffham, "The Internet of Things: Impact and implications for health care delivery", *J. Med. Internet Res.,* vol. 22, no. 11, p. e20135, 2020. [http://dx.doi.org/10.2196/20135] [PMID: 33170132]

[8] A. Mohsin, and S.S. Yellampalli, "IoT based cold chain logistics monitoring", *2017 IEEE International Conference on Power, Control, Signals and Instrumentation Engineering (ICPCSI),* IEEE., pp. 1971-1974, 2017. [http://dx.doi.org/10.1109/ICPCSI.2017.8392059]

[9] S. Aich, S. Chakraborty, M. Sain, H.I. Lee, and H.C. Kim, "A review on benefits of IoT integrated blockchain based supply chain management implementations across different sectors with case study", *In 2019 21st international conference on advanced communication technology,* pp. 138-141, 2019. [http://dx.doi.org/10.23919/ICACT.2019.8701910]

[10] A. Premkumar, and C. Srimathi, "Application of blockchain and iot towards pharmaceutical industry", *In 2020 6th International Conference on Advanced Computing and Communication Systems,* pp. 729-733, 2020. [http://dx.doi.org/10.1109/ICACCS48705.2020.9074264]

[11] S. Bhalerao, S. Agarwal, S. Borkar, S. Anekar, N. Kulkarni, and S. Bhagwat, "Supply chain management using blockchain", In: *In 2019 international conference on intelligent sustainable systems*, 2019, pp. 456-459. [http://dx.doi.org/10.1109/ISS1.2019.8908031]

[12] W. Sun, "Application of blockchain technology in the supply chain finance", *In 2022 7th International Conference on Cloud Computing and Big Data Analytics,* pp. 205-209, 2022. [http://dx.doi.org/10.1109/ICCCBDA55098.2022.9778290]

[13] B. Musigmann, H. von der Gracht, and E. Hartmann, "Blockchain technology in logistics and supply chain management—A bibliometric literature review from 2016 to January 2020", *IEEE Trans. Eng. Manage.,* vol. 67, no. 4, pp. 988-1007, 2020. [http://dx.doi.org/10.1109/TEM.2020.2980733]

[14] P. Ratta, A. Kaur, S. Sharma, M. Shabaz, and G. Dhiman, "Application of blockchain and internet of things in healthcare and medical sector: Applications, challenges, and future perspectives", *J. Food Qual.,* vol. 2021, pp. 1-20, 2021. [http://dx.doi.org/10.1155/2021/7608296]

[15] S.K. Das, N. Dey, R. González Crespo, and E. Herrera-Viedma, "A non-linear multi-objective technique for hybrid peer-to-peer communication", *Inf. Sci.,* vol. 629, pp. 413-439, 2023. [http://dx.doi.org/10.1016/j.ins.2023.01.117]

[16] C. K. M. Lee, M. N. Cheng, and C. K. Ng, "IoT-based asset management system for healthcare-related industries", *Int. J. Eng. Bus. Manag.,* pp. 7-19, 2015.

[17] A.K. Dass, and S.K. Das, "An optimal data communication system in wireless sensor network using artificial neural network", In: *Constraint Decision-Making Systems in Engineering.* IGI Global, 2023, pp. 120-134.

[http://dx.doi.org/10.4018/978-1-6684-7343-6.ch007]

[18] P. Saindane, Y. Jethani, P. Mahtani, C. Rohra, and P. Lund, "Blockchain: A solution for improved traceability with reduced counterfeits in supply chain of drugs", *2020 International Conference on Electrotechnical Complexes and Systems (ICOECS),* IEEE., pp. 1-5, 2020. [http://dx.doi.org/10.1109/ICOECS50468.2020.9278412]

[19] Y. Raj, and B. Sowmiya, "Study on supply chain management using blockchain technology", *In 2021 6th International Conference on Inventive Computation Technologies,* pp. 1243-1247, 2021. [http://dx.doi.org/10.1109/ICICT50816.2021.9358768]

[20] A. Sharma, J. Kaur, and I. Singh, "Internet of things (IoT) in pharmaceutical manufacturing, warehousing, and supply chain management", *SN Computer Science,* vol. 1, no. 4, p. 232, 2020. [http://dx.doi.org/10.1007/s42979-020-00248-2]

[21] T. Bocek, B.B. Rodrigues, T. Strasser, and B. Stiller, *Blockchains everywhere-a use-case of blockchains in the pharma supply-chain. In 2017 IFIP/IEEE symposium on integrated network and service management (IM).* IEEE, 2017, pp. 772-777.

[22] V. Sudha, R. Kalaiselvi, and P. Shanmughasundaram, "Blockchain based solution to improve the supply chain management in indian agriculture", *2021 International Conference on Artificial Intelligence and Smart Systems (ICAIS),* IEEE., pp. 1289-1292, 2021. [http://dx.doi.org/10.1109/ICAIS50930.2021.9395867]

[23] U. Agarwal, V. Rishiwal, S. Tanwar, R. Chaudhary, G. Sharma, P.N. Bokoro, and R. Sharma, "Blockchain technology for secure supply chain management: A comprehensive review", *IEEE Access,* vol. 10, pp. 85493-85517, 2022. [http://dx.doi.org/10.1109/ACCESS.2022.3194319]

[24] T. Abdelgalil, V. Manolas, L. Maglaras, I. Kantzavelou, and M.A. Ferrag, "Blockchain technology: A case study in supply chain management", *2021 Third IEEE International Conference on Trust, Privacy and Security in Intelligent Systems and Applications (TPS-ISA),* pp. 331-339, 2021. [http://dx.doi.org/10.1109/TPSISA52974.2021.00036]

[25] D. Sathya, S. Nithyaroopa, D. Jagadeesan, and I.J. Jacob, "Block-chain technology for food supply chains", *2021 Third International Conference on Intelligent Communication Technologies and Virtual Mobile Networks (ICICV),* IEEE., pp. 212-219, 2021.

[26] S.T. Sadasivuni, Y. Zhang, G. Thejasree, and R. Maniyeri, "Table of contents", *2019 IEEE International Conference on Intelligent Systems and Green Technology (ICISGT),* pp. 5-53, 2019.Visakhapatnam, India

[27] M. Hader, A. Elmhamedi, and A. Abouabdellah, "Blockchain technology in supply chain management and loyalty programs: toward blockchain implementation in retail market", *In 2020 IEEE 13th International Colloquium of Logistics and Supply Chain Management,* pp. 1-6, 2020. [http://dx.doi.org/10.1109/LOGISTIQUA49782.2020.9353879]

[28] S. Darbari, *Blockchain-based solutions for privacy in the internet of things smart environment.,* 2021.

[29] S. Majumder, and N. Dey, *AI-empowered knowledge management.* Springer, 2022. [http://dx.doi.org/10.1007/978-981-19-0316-8]

[30] S.K. Das, S. Samal, P. Ranjan, and S.P. Das, "An efficient data mining technique for an intrusion detection system in network", In: *Constraint Decision-Making Systems in Engineering.* IGI Global, 2023, pp. 1-17. [http://dx.doi.org/10.4018/978-1-6684-7343-6.ch001]

[31] S. K. Das, "Smart design and its applications: Challenges and techniques", *Nature-Inspired Computing for Smart Application Design,* pp. 1-6, 2021.

[32] S.T. Jyothy, and M. Sarvagya, "Supply Chain Management in Pharmaceutical Industry Using IOT", *2022 IEEE North Karnataka Subsection Flagship International Conference (NKCon),* IEEE., pp. 1-7, 2022.

CHAPTER 9

Attrition in IT Sector: Psychology Behind the Scene

Abhisek Sarkar[1,*]

[1] *Dinabandhu Andrews Institute of Technology and Management, kolkata, India*

Abstract: Attrition is often defined as "a reduction in the number of employees as a result of retirement, resignation or death" and also as "the rate of shrinkage in size or number". But the scenario is not so simple. We should always consider premature retirement, sudden resignation and premature death, including suicide. The real scenario is employees do leave, either because they expect extra money, dislike the working environment, get rough behavior, non-cooperation from their coworkers, need a change, or because their spouse gets a more robust chance in another region. Retention is more economical than going for brand spanking new recruitment whatsoever. Organizations should have a good retention strategy to retain their valuable employees. Employee turnover may be viewed as the outcome of unmet expectations and gaps between fundamental employee demands. Employees may simply resign under a few unfavourable conditions, but more crucially, "people depart before they leave", according to the psychology of disengagement. It may be iterated that as they become older, their contribution gradually decreases, much like a slowly fading memory. This text presents a holistic view of attrition and retention of employees based on psychological aspects in this cut-throat competitive environment in India. Biology has a little role in management, though one cannot ignore biology in psychology. In broader terms, attrition is somehow related to psychology, and psychology and physiology are two sides of a coin. A new trend is to relate psychology with physiology to reduce attrition.

Keywords: Attrition, Employee recognition, Meta analysis, Psychology, Physiology, Retention.

INTRODUCTION

The natural process through which employees quit the workforce is known as employee attrition. Attrition is the common occurrence of employees leaving a company or organisation voluntarily (or involuntarily) for reasons related to the environment and culture of the company that can be both professional and personal. Attrition in a corporation is typically calculated using a metric called

* **Corresponding author Abhisek Sarkar:** Dinabandhu Andrews Institute of Technology and Management, kolkata, India; E-mail: abhisar0210@gmail.com

Santosh Kumar Das, Soumi Majumder and Nilanjan Dey (Eds.)

attrition rate, which counts the number of workers leaving the organisation who may be voluntarily resigning or laid off by the company. Churn rate and turnover are other names for attrition rate. However, compared to attrition, turnover has a tendency to be more abrupt. Internal migration in a company may or may not be a factor in overall attrition, but it can affect the attrition of a specific unit. The rate at which employees leave a company is referred to as the attrition rate. The formula shown in Eq. (1) can be used to compute this rate.

Attrition Rate (%) = (Number of Separations/Average Number of Employees) X 100 ....(1)

The average staff count takes into account both new hires and departures within that specific time period. It provides information on the number of workers quitting the company at any particular time. It is a significant factor because businesses must get ready to begin hiring for crucial roles that cannot be left open. Any firm will inevitably experience attrition. An employee will eventually desire to quit his or her present organisation, whether for personal or professional reasons. It may also include a harsh work environment, a lack of professional advancement, or a loss of faith in the market value of the business, among others. Another issue that frequently contributes to employee's reasons to leave is poor leadership. Notably, all these reasons are directly or indirectly related to psychology.

However, attrition only becomes a problem when it exceeds a certain threshold. For instance, minor employee attrition may be detrimental to the company's diversity efforts, or a major leadership gap in the organisation can result from the departure of top leaders. If employees believe they are being exploited or devalued, employee turnover is more likely. The workplace has a significant impact on employees' lives. If they do not feel protected or appreciated, they may feel compelled to leave. Additionally, if managers do not treat all employees fairly, it will lead to excessive employee turnover.

Basic management practices, such as staff perks and incentives, can also have an impact on the environment. If any of these factors are not correctly and successfully applied, the desired outcomes (motivated employees) can not be achieved.

Additionally, their hiring practices and layoff policies have an impact on the workplace. Due to departmental policies or coworkers' actions, many individuals quit their existing jobs and accept new ones within the same organisation.

Factors Leading to Employee Attrition

There are a number of reasons why workers think about leaving their current employer. Following are some of the primary and general causes of employee attrition:

- Better income and job prospects outside the company.
- Low pay or lack of employee recognition.
- Improper work-life balance.
- Rude behaviour on the part of peers and bosses, which results in poor team management and demoralisation.
- A lower work-life balance and stagnant career progress.
- Poor and inadequate working circumstances that sap motivation.
- Employees' premature retirement or fatal accidents while working.
- Women have many roles to perform throughout their life. They work nonstop, from performing their family responsibilities to achieving their goals. And if they are not content and at ease, they just quit their work. Some choose to leave the company outright rather than take maternity leave. These elements contribute to a higher turnover of women.

Administrators must first comprehend the psychology of praise and gratitude in order to improve employee retention. They will gain a better understanding of what motivates their staff if they do this. In essence, appreciation and gratitude are intrinsic motivational concepts. There are two categories of motivation in psychology: intrinsic and extrinsic. When someone is motivated internally, they operate without expecting any external rewards. A person who is intrinsically motivated will work on a task because they find enjoyment in it and gain internal fulfillment from it. Extrinsic motivation, on the other hand, is the act of performing something in order to obtain an external reward.

According to a study on the psychological impacts of recognition and acknowledgment at work, these two factors are crucial for both employee motivation and organisational performance. Employers can lower their employee turnover rate by frequently showing appreciation to their staff. This study also showed that lower staff turnover and more employee commitment result from treating employees with respect and appreciation for their contributions at work.

The Science Behind the Motivation

All types of motivation are greatly influenced by social psychology, cognitive psychology, cognitive neuroscience, and psychometrics. Motivation is a factor in almost every aspect of human behaviour. Undoubtedly, a person's motivational state has an impact on how they make decisions.

In recent years, Braver *et al.* have realized the significance of a more comprehensive and interdisciplinary approach to the study of motivation [1]. Multiple approaches like behavioral experiments, neuroimaging, meta-analysis, statistical simulation/computational modeling, longitudinal data analysis, and network analysis may be helpful to find out the exact reason behind the motivation.

Dopamine is a neurotransmitter that contributes to the study of motivation. The chemical signal of dopamine is transmitted from one neuron to the next, where it interacts with different receptors in the synapse between those two neurons. Which pathway dopamine uses is important, especially for motivation. The most significant reward pathway in the brain is the mesolimbic pathway, which originates from the centre of the brain and branches to several locations, including the cerebral cortex. Dopamine is involved in the regulation of motivation and reward and plays a role in the experience of pleasure and satisfaction. Low levels of dopamine are associated with reduced motivation and a lack of interest in rewarding activities while increasing dopamine levels can increase motivation and improve mood. Serotonin is involved in the regulation of emotions, including the regulation of negative emotions such as frustration. Low levels of serotonin are associated with increased irritability and aggression while increasing serotonin levels through medication or other means can help reduce frustration and improve mood. According to researchers Aberman and John Salamone of the University of Connecticut, "Low levels of dopamine make people and other animals less likely to work for things, so it has more to do with motivation and cost/benefit analyses than pleasure itself" [2]. Dopamine levels increase when one anticipates an important event because he/she is motivated. The brain can be made to thrive on dopamine surges brought on by pleasurable experiences. The brain takes care of the rest as one establishes the dopamine environment.

Relation Between Attrition and Dopamine

We feel satisfied when dopamine levels in the brain are normal. Our desire to work eventually rises. We accomplish things, make thoughtful plans, and become quick learners. We have a strong sense of purpose and become enthusiastic about life. Dopamine levels that are in the normal range can also make us more outgoing and sociable. This "feel good" neurotransmitter also aids in enhancing our capacity for empathy, which increases our willingness to accommodate others' needs. Dopamine can promote creativity as well. All of these characteristics contribute to the dopamine-induced pleasure. However, when the brain is not getting enough dopamine, our motivation declines, our emotions alter, and our conduct may significantly deteriorate.

We may experience fatigue and restlessness due to low dopamine levels, commonly referred to as "dopamine insufficiency". Low amounts of dopamine might make us feel demotivated, melancholic, and nervous instead of full of life. We become less focused. Our sleep patterns deteriorate. We start to feel mentally clouded. When the brain does not make enough dopamine, even our body finds it difficult to feel alive. Motivation, reinforcement, and reward can be considered as a cycle. Sometimes, just anticipating something might increase dopamine levels if employees associate it with pleasure.

Dopamine plays important roles in pleasure and rewards, but it also serves the following other purposes:

- Motor functions
- Mood regulation
- Executive functioning
- Memory and focus
- Sleep
- Stress response
- Digestion and blood flow
- Anxiety
- Addiction
- Behavioral disturbances
- Cognitive disorders
- Diseases (Such as Parkinson's)
- Fatigue
- Hormonal imbalances
- Obesity
- Pain

So, the factor for attrition may be noted due to lack of focus, sleep and digestion, which in turn compromises health. Above all, low dopamine can promote stress.

Attrition in the IT Industry

The "Talent Exodus Report", published by the IT staffing firm TeamLease Digital, claims that IT companies are suffering dysfunctional turnover, in which the best employees of businesses leave at increased rates of their own volition [3]. The IT business is currently experiencing "Brain Drain", and according to a survey, people are reevaluating their jobs and seeking flexibility, as reported by New Indian Express on November 22, 2022. It also reported that the $227 billion IT sector of the country currently employs over 5 million people. While Infosys, an IT company, recorded the highest attrition rate in the first quarter at 28.4%, and

Wipro's was 23.3%. With regard to contract staffing, in India, attrition canincrease from 49% in the previous fiscal year to 55% in the current fiscal year.

As corporations deal with cost overloads and margin challenges, record-high attrition rates are compounding the problems being faced by the Indian IT sector. According to a report of Business Today, Sep 02, 2022, the top three Indian IT behemoths, Tata Consultancy Services (TCS), Infosys, and Wipro, increased their net hirings to over 50,000 in the first quarter of the fiscal year 2022–23 due to high attrition [4].

According to Sunil Chemmankotil, CEO of TeamLease Digital, talent is moving due to company rules, compensation, and work flexibility as non-technology industries have increased their demand [5]. "In the upcoming years, tech talent in non-tech industries will rise by 3 folds, creating over 1 million additional tech employment by 2025. In the near future, this will result in more IT talent leaving traditional firms." He also added, "Although the IT sector is expanding, it is getting harder to keep personnel. According to a survey, 57% of IT professionals said that they would not contemplate going back to the IT services industry in the future. The main draw for employees in new jobs, despite the customary demand for pay increases and other perks, is "Great Reflection" on internal policies and outside factors that employers should review in light of the significant shifts in employees' attitudes toward their jobs and overall well-being."

It is commonly recognized that physiology and psychology are related. Arnold M. Mordkoff, in his article "The Relationship between Psychological and Physiological Response to Stress", highlighted the association between physiological and psychological responses to stress and the circumstances under which the degree of relationship could be adjusted [6]. In spite of the experimental modifications added to the experiment, the degree of the link between psychological and physiological reaction, as measured by multiple regression analyses of the responses of each segment, was impressively consistent. Similar to continuous monitoring of physiological reactions, an intra-individual methodology was employed to gather reports of psychological experience during a stressor film stimulus.

Geoff Thompson wrote about the need for sleep for maintaining daily routines in his book, Stress Buster [7]. Overwork, stress, and overtime may cause changes in the biological clock. Most people are aware that they naturally fluctuate between various states of alertness and sleepiness throughout the day, but why are these patterns present? The circadian biological clock and the sleep/wake equilibrium are the two bodily mechanisms that control sleep. Physiological activity may become resistant to sleep as a result of a change in the biological clock. Lack of inventiveness, dyspepsia, loss of memory, and lack of concentration are the ultimate effects.

Literature Review

Mohlala, J. *et al.* (2012) investigated employee retention in their SA Journal of Human Resource Management/SA Tydskrif vir Mens like hulpbronbestuur within the Information Technology Division of a South African Bank [8]. They noticed that the ability to retain information technology specialists depends on both situational obstacles (which involve the interaction of organisational and industry-specific issues) and people-specific issues in the field. Issues of the information technology personnel were due to a lack of passion in a career for life, unattractive technology, and improper rewards and recognition. The difficulties that particular information technology workers face have an indirect impact on retention.

Kehr (2004), in his research paper "Integrating Implicit Motives, Explicit Motives, and Perceived Abilities: The Compensatory Model of Work Motivation and Volition," discussed that three variables—power, achievement, and affiliation—can be used to classify the implicit retention aspects that influence spontaneous, expressive, and enjoyable behaviour [9]. Domination and social control are two definitions of power. Affiliation describes the establishing and strengthening of social bonds, whereas achievement refers to the attainment of one's own standards of excellence. Although implicit and explicit retention factors are related to various facets of the individual, both are significant behavioural predictors.

Fitz-enz (1990) noted that a number of factors work together to keep people in a company rather than just one element having an impact on employee retention [10]. Related issues, including income and benefits, development opportunities, job security, workplace culture, *etc.*, need to be taken into consideration by the management. Fitz-enz's 1990) theory (1990) that job change motivation is multimodal and that no single factor can explain it was supported by a study on turnover conducted in New Zealand by Boxall *et al.* in 2003 [11]. According to Boxall *et al.* (2003), work-related accidents or illnesses, unhappiness with coworkers, commuting, unrealistic expectations from the job, excessive work demands, lack of promotion elsewhere, non-redressal of grievances, work methods, lack of job security, inadequate pay, change of career, difficult relationships with the supervisor, work-life demands, lack of training opportunities, non-recognition of employee merit, and more interesting work elsewhere may cause employees to leave their jobs.

Walker (2001), in his article "Zero Defections? Human Resource Planning", identified seven crucial factors that can help retain employees [12]:

- pay and recognition for the work done;
- challenging work;
- opportunities for advancement and learning;
- an inviting environment within the organisation;
- good relationships with coworkers;
- a healthy balance between professional and personal life; and
- good communications.

Trevor *et al.* (1997) suggested that it has been established that increase in compensation have a negative effect on turnover [13]. According to Davies *et al.* (2001), although every business pays its top employees, very few of them do so purposefully [14]. According to them, "Salary and benefit policies are not being used strategically inside the business to promote morale, reduce turnover, and meet establishment-wide goals."

Hyman *et al.* (2003), in their empirical research, discovered that the intrusion of work obligations into personal life caused employees' stress levels to rise and their emotional stamina to wane [15].

Johnson (1981) believed that turnover was a significant issue that had a negative impact on the quality of goods and services and resulted in high replacement and recruitment costs [16].

Bogdanowicz and Bailey (2002) stated that, for a business to survive, it is imperative to retain valuable employees or functional employees [17].

Messmer (2000) identified that investment in employee training and career development is one of the key determinants in employee retention [18]. An organisation will always spend money on training and development for employees who they know will produce an equal or greater return on their investment.

According to Aliya and Fariduddin (2003), there are differences in the elements that lead to positive and negative views [19]. Retention of employees is greatly influenced by attitudes.

Griffeth *et al.* (2000) noticed that organisational commitment was a greater predictor of turnover than even work satisfaction when they identified job satisfaction as a potential precursor of turnover [20].

Glance *et al.* (1997) justified the correlation between turnover and production [21]. They asserted that higher production was unquestionably connected with reduced turnover rates.

Attrition is caused by problems such as stress, sexual and racial abuse, and job discontent, as stated by Walletwatch (2003), Cacanas (2004), Singh (2005a), Witt *et al.* (2005), Rose and Wright (2005), and Houlihan (2004) [22 - 27].

Dundon *et al.* (2004) investigated and identified that effective orientation is one of the mechanisms that lowers turnover, especially during the first few months of work [28].

Mano-Negrin and Kirschenbaum (2000) assert that organisational size has an impact on turnover. They highlight that the main ways in which organisational size affects turnover are through salary rates and career advancement opportunities [29].

Age was noted by Wright and Bonett in 2002 as a factor influencing job performance [30]. Non-financial benefits, such as accomplishment acknowledgment, were discovered by Cronin and Becherer in 1999 and Fort and Voltero in 2004 [31, 32].

Swift and Greenberg (2012), Corning and Malofeeva (2004), and Reis and Brown (1999) discussedthat attrition, also known as premature discontinuation, premature termination, and premature discontinuation of treatment, has been noted in recent literature as a frequent occurrence that poses a significant barrier to the provision of effective psychotherapy [33 - 35]. It was also supported by Barrett, Chua, Crits-Christoph, Gibbons, and Thompson (2008) [36]. Additionally, it appears that attrition has an effect on more than just client outcome, as noted by Reis and Brown (1999) and Swift and Greenberg (2012). According to Swift and Greenberg (2012), society as a whole is burdened by attrition and is forced to endure the "continuous impairment" of its members.

Action of Dopamine and Serotonin in Controlling Emotions

Both dopamine and serotonin affect mood and concentration and relay messages between neurons, but they also have other important purposes. For instance, dopamine transmits messages between neurons that regulate balance and coordination in the body.

Numerous behaviours are controlled by this neurotransmitter, which also functions in the brain's pleasure and reward region. Dopamine levels in the brain can rise as a result of ingesting particular foods, using illicit drugs, and engaging in certain behaviours like gambling.

Increased dopamine levels can result in happiness, bliss, and improved motivation and focus. As a result, some people may develop an addiction to substances and activities that enhance dopamine. Serotonin and dopamine are neurotransmitters that have favourable connections.

They are also neuromodulators, which means that, in contrast to other neurotransmitters, they can interact with several neurons both close to and far from the source of dopamine or serotonin production. Neurotransmitters do not operate on their own. They interact and have an impact on one another as the body's delicate chemical equilibrium is maintained. Both physically and functionally, the serotonin and dopamine systems are closely related.

However, serotonin may occasionally prevent the synthesis of dopamine, which means that a lack of serotonin may cause an excess of dopamine to be produced. Due to the part dopamine plays in behaviour that seeks rewards, this can result in impulsive conduct.

Serotonin and dopamine have opposing impacts on hunger as well. Low amounts of dopamine can increase hunger, whilst serotonin can inhibit it. Additionally, serotonin controls impulsive behaviour. Dopamine, on the other hand, increases impulsivity.

According to researchers Barrett *et al.* (2008), Hatchett and Park (2003), and Swift and Greenberg, (2012), the disparate results in independent attrition studies are the result of different ways that academics have operationalized the concept of attrition [37 - 39]. People who prematurely stop therapy are those who attend fewer than a certain number of therapy sessions, according to one well-known model used to comprehend the concept of attrition, according to Hansen *et al.* (2002) [40]. This model is based on the dose-effect relationship theory, where the dosc is determined by the number of sessions a client attends and the effect is determined by the proportion of clients who experience improved outcomes, which is also known as 'the normalised probability of improvement for one patient' (Howard *et al.* (1986)) [41]. As stated by Hansen *et al.* (2002), "Dose-response approaches have been widely accepted in medical research and have more recently been adopted in psychotherapy research" [42]. According to the dose-effect relationship, "clients must attend a minimum number of sessions to demonstrate improvement in therapy" (Swift and Greenberg, 2012, p. 548). Since it is often believed that more therapy is better, 156 studies over the past 65 years were examined, and 100 of them "suggested a positive link between therapy time and outcome" (Anderson & Lambert, 2001, p. 876) [43].

Dopamine, Serotonin and Sleep

Both serotonin and dopamine have an impact on the sleep-wake cycle. The pineal gland in the brain, which has receptors for both neurotransmitters, controls this cycle. The hormone melatonin, which induces sleepiness, is secreted by the pineal gland in response to signals from the eyes about light and darkness. According to a study by Nora D. *et al.*, dopamine significantly affects wakefulness by reducing norepinephrine production [44].

Researchers discovered that dopamine might induce rapid eye movement (REM) sleep in rats, according to a study conducted in 2022. People who have narcolepsy may also be affected by the same mechanism, as an increase in dopamine may contribute to uncontrollable daytime sleep episodes.

Both wakefulness and drowsiness are aided by serotonin. To create melatonin, a human needs serotonin. But serotonin also inhibits REM sleep, which could result in sleep disorders.

Dopamine, Serotonin and Digestion

The gut is where the body produces the majority of serotonin. The small intestine is stimulated by serotonin to contract, which helps food pass through the digestive system.

When a person ingests anything to which they are allergic or that includes hazardous microorganisms, more serotonin is released. Faster contractions as a result of this may produce vomiting or diarrhea.

On the other hand, a low level of serotonin in the intestines may result in constipation.

Dopamine, like serotonin, aids in the movement of food through the digestive system. Additionally, it preserves the mucosal lining of the gut, which may help avoid peptic ulcers, and aids in controlling the synthesis of insulin by the pancreas.

Dopamine, Serotonin and Mental Health

Dopamine and serotonin both have an effect on mental health.

Dopamine

Although low levels of dopamine may not directly cause depression, they may create some symptoms that medical professionals believe the disorder is associated with.

These signs may manifest as:

- a lack of drive.
- having trouble concentrating and feeling sad and helpless.
- loss of enthusiasm for once-fun activities.

Serotonin

Low serotonin levels have been linked to mood disorders like anxiety and sadness by researchers.

Many elements other than biochemistry also play a role in depression, including:

- genealogy and genetics.
- surroundings, stress levels, and extra medical concerns.

A person's chance of experiencing depression may rise if their serotonin levels are low.

A higher risk of addiction may also be associated with mental health problems. According to research, those who suffer from mental illnesses may utilise mind-altering medications more frequently. According to some researchers, people who have been diagnosed with a mental health condition like schizophrenia or attention deficit hyperactivity disorder are more likely to use illegal drugs to "self-treat" their condition.

Attrition Rate in Indian and India-based IT/ITeS firms

The rate of attrition in the Indian IT services sector has been extremely high and has been negatively affecting the sector since the quarter ending in June 2021 (Q1 FY22), when 4.89 lakh people left the top eight organisations, including Tata Consultancy Services, Cognizant, Infosys, and Wipro.

The approximately 500,000 attrition-related talent moves, according to Kamal Karanth, co-founder of Xpheno, represent a loss of expertise, experience, and upskilling investments that businesses made in this personnel [45]. Businesses have a lot to explain and defend in their P&Ls after factoring in the cost of hiring more employees, raising salaries, and investing in retraining. (The Hindu Newspaper, dated August 10, 2022) [46].

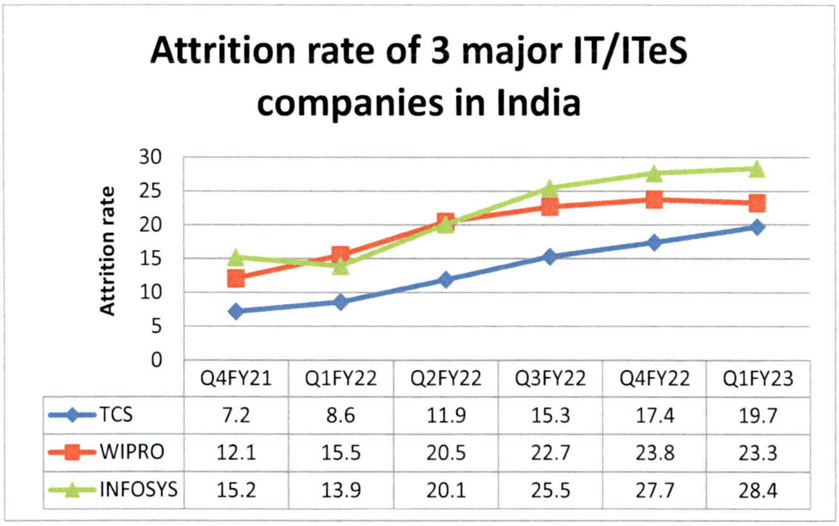

(Source: Business Today survey and company results) [47]

The term "attrition rate" refers to a metric used to track employee losses over time. In the first quarter of the fiscal year 2022–23, Infosys had the highest attrition rate among Indian IT leaders at 28.4%. The attrition rate at Tech Mahindra was reported to be 22%, while the attrition rate at Wipro was 23.3%. Tech Mahindra's Managing Director and Chief Executive Officer, CP Gurnani, reasoned that the industry's quick expansions might be to blame for the high turnover rates. He stated, "In general, there is shock when an industry experiences a quick increase, with attrition rate running at 23-24%, and I think that shock stage is now abating." The fact that there are too many people employed in the Indian IT sector was also emphasised by T.V. Mohandas Pai, a former director of Infosys. There is an excess of individuals in the system, he told Business Today. All of them desire to work for large corporations, develop their careers, and receive quality education. Because of the enormous amount of talent action they clocked over the past five quarters, Indian IT service bellwethers will always be remembered for them. Managing huge attrition spikes while achieving record-high net workforce additions is no simple task.

While Cognizant routinely reported over 30% attrition over the previous five quarters, this number doubled for peers TCS, Infosys, and HCL Tech between Q1 FY22 and Q1 FY23, and increased significantly for Wipro over the same period.

CONCLUSION

Attrition, or the process of employees leaving an organization, can be linked to various psychological factors. Some possible links include job satisfaction, organizational commitment, stress, burnout, and work-life balance. Individuals who experience high levels of stress or burnout, low job satisfaction, or poor work-life balance may be more likely to leave their jobs. Additionally, employees who feel less committed to their organization may be more likely to consider leaving. Understanding these psychological factors and addressing them through effective organizational practices can help reduce attrition and improve employee retention. One of the key factors that psychology highlights is the importance of employee engagement. Engaged employees are more likely to stay with an organisation as they feel invested in their work and are motivated to contribute to the success of the company. Factors that contribute to engagement include feeling valued by the organization, having a clear sense of purpose and direction, and having opportunities for growth and development.

Research suggests that imbalances in serotonin and dopamine can contribute to various psychological disorders, such as depression, anxiety, and addiction. For example, low levels of serotonin are linked to depression, while disruptions in dopamine function are associated with addiction and other impulsive behaviors. Simply said, a happy hormone is a hormone (or, technically, in some situations, a neurotransmitter) that makes us feel good. So simply, if we relate the good or happy hormone with unsatisfied or dissatisfied employees, attrition may be reduced. A new concept of DOSE has emerged. D for Dopamine, which we can relate with Drive, O for Oxytosin, which may be related with Oneness, S for Serotonin, which may be connected with Stability and E for Endorphin, which may be related with Ease. In addition, some medications that affect one of these neurotransmitters can also affect the other. For example, Selective Serotonin Reuptake Inhibitors (SSRIs), which are commonly used to treat depression and anxiety, can indirectly affect dopamine levels.

Overall, serotonin and dopamine are two important neurotransmitters that work together to regulate various aspects of psychological functioning, and imbalances in either neurotransmitter can have significant psychological effects on attrition.

REFERENCES

[1] "A M mordkoff, The relationship between psychological and physiological response to stress", *Psychosom Med.,* vol. 26, pp. 135-150, 1946.

[2] J.E. Aberman, and J.D. Salamone, "Nucleus accumbens dopamine depletions make rats more sensitive to high ratio requirements but do not impair primary food reinforcement", *Neuroscience,* vol. 92, no. 2, pp. 545-552, 1999.
[http://dx.doi.org/10.1016/S0306-4522(99)00004-4] [PMID: 10408603]

[3] The Indian Express, Over 2 million employees expected to leave their jobs by 2025, finds talent exodus report By: Careers Desk , Edited by Deeksha Teri New Delhi | Updated: September 30, 2022 09:37 IST, 2022.

[4] "Business Today, "Why Infosys, TCS, Wipro, and other Indian IT giants are facing record-high attrition rates", by Aakanksha Chaturvedi", 2024",

[5] BW People, "22 Lakh Indian IT Professional Will Leave Job By 2025", *TeamLease Digital Report,* 2022.

[6] A.M. Mordkoff, "The relationship between psychological and physiological response to stress", *Psychosom. Med.,* vol. 26, no. 2, pp. 135-150, 1964. [http://dx.doi.org/10.1097/00006842-196403000-00005] [PMID: 14132975]

[7] Geoff Thompson, and Stress Buster, *How to Stop Stress from Killing You, p177.,* 2021.

[8] J. Mohlala, G.A. Goldman, and X. Goosen, "Employee retention within the information technology division of a south african bank", *SA Journal of Human Resource Management/SA Tydskrif vir Menslikehulpbronbestuur,,* vol. 10, no. 2, 2012.

[9] H.M. Kehr, "Integrating implicit motives, explicit motives, and perceived abilities: the compensatory model of work motivation and volition", *Acad. Manage. Rev.,* vol. 29, no. 3, p. 479, 2004. [http://dx.doi.org/10.2307/20159055]

[10] J. Fitz-enz, "Getting and keeping good employees", *In Personnel,* vol. 67, pp. 25-29, 1990.

[11] P. Boxall, K. Macky, and E. Rasmussen, "Labour turnover and retention in new zealand: the causes and consequences of leaving and staying with employers", *Asia Pac. J. Hum. Resour.,* vol. 41, no. 2, pp. 196-214, 2003. [http://dx.doi.org/10.1177/10384111030412006]

[12] J.W. Walker, "Zero defections?", *Human Resource Planning,* vol. 24, pp. 6-8, 2001.

[13] C.O. Trevor, B. Gerhart, and J.W. Boudreau, "Voluntary turnover and job performance: Curvilinearity and the moderating influences of salary growth and promotions", *J. Appl. Psychol.,* vol. 82, no. 1, pp. 44-61, 1997. [http://dx.doi.org/10.1037/0021-9010.82.1.44]

[14] D. Davies, R. Taylor, and L. Savery, "The role of appraisal, remuneration and training in improving staff relations in the Western Australian accommodation industry: a comparative study", *J. Eur. Ind. Train.,* vol. 25, no. 7, pp. 366-373, 2001. [http://dx.doi.org/10.1108/EUM0000000005837]

[15] Jeffrey Hyman, Chris Baldry, Dora Scholarios, and Dirk Bunzel, *Work-Life Imbalance in the New Service Sector Economy.,* 2003.

[16] K. Johnson, "Towards an understanding of labour turnover, service industries review,", *African Journal of Business Management,* vol. 1, no. 3, pp. 49-54, 2007.

[17] M.S. Bogdanowicz, and E.K. Bailey, "The value of knowledge and the values of the new knowledge worker: Generation X in the new economy", *J. Eur. Ind. Train.,* vol. 26, no. 2/3/4, pp. 125-129, 2002. [http://dx.doi.org/10.1108/03090590210422003]

[18] M. Messmer, "Orientations programs can be key to employee retention", *Strategic Finance,* vol. 81, pp. 12-15, 2000.

[19] Aliya Ahmad and Shadab Fariduddin, "The motivation to work, Journal of business and organization development", *The International Journal of Business and Management,* vol. 9, no. 11, pp. 298-304, 2021.

[20] R.W. Griffeth, P.W. Hom, and S. Gaertner, "A meta-analysis of antecedents and correlates of employee turnover: Update, moderator tests, and research implications for the next millennium", *J. Manage.,* vol. 26, no. 3, pp. 463-488, 2000.

[http://dx.doi.org/10.1177/014920630002600305]

[21] N.S. Glance, T. Hogg, and B.A. Huberman, "Training and turnover in the evolution of organizations", *Organ. Sci.,* vol. 8, no. 1, pp. 84-96, 1997. [http://dx.doi.org/10.1287/orsc.8.1.84]

[22] "Walletwatch 'call centre attrition puts HR managers to test",

[23] Z. Cacanas, "Passage to India", *Human Resources,* no. March, pp. 46-50, 2004.

[24] H. Singh, "(2005) "Is the BPO iceberg melting under attrition heat?" The Economic Times, 10th February. ", *International Journal of Marketing and Human Resource Management,* vol. 1, no. 1, pp. 18-27, 2010.

[25] L.A. Witt, M.C. Andrews, and D.S. Carlson, "When conscientiousness isn't enough: Emotional exhaustion and performance among call center customer service representatives", *J. Manage.,* vol. 30, no. 1, pp. 149-160, 2004. [http://dx.doi.org/10.1016/j.jm.2003.01.007]

[26] E. Rose, and G. Wright, "Satisfaction and dimensions of control among call centre customer service representatives", *Int. J. Hum. Resour. Manage.,* vol. 16, no. 1, pp. 136-160, 2005. [http://dx.doi.org/10.1080/0958519042000295000]

[27] M. Houlihan, "Tensions and variations in call centre management strategies", In: *Call Centres and Human Resource Management: A Cross National Perspective.,* S. Deery, N. Kinnie, Eds., Palgrave Macmillan: Basingstoke, 2004, pp. 75-101. [http://dx.doi.org/10.1057/9780230288805_4]

[28] T. Dundon, A. Wilkinson, M. Marchington, and P. Ackers, "The meanings and purpose of employee voice", *Int. J. Hum. Resour. Manage.,* vol. 15, no. 6, pp. 1149-1170, 2004. [http://dx.doi.org/10.1080/0958519041000167733591]

[29] R. Mano-Negrin, and A. Kirschenbaum, "A Spousal interdependence in turnover decisions", *J. Fam. Econ. Issues,* vol. 21, no. 2, pp. 97-122, 2000. [http://dx.doi.org/10.1023/A:1009421901790]

[30] T.A. Wright, and D.G. Bonett, "The moderating effects of employee tenure on the relation between organizational commitment and job performance: A meta-analysis", *J. Appl. Psychol.,* vol. 87, no. 6, pp. 1183-1190, 2002. [http://dx.doi.org/10.1037/0021-9010.87.6.1183] [PMID: 12558224]

[31] S.N. Cronin, and D. Becherer, "Recognition of staff nurse job performance and achievements: Staff and manager perceptions", *J. Nurs. Adm.,* vol. 29, no. 1, pp. 26-31, 1999. [http://dx.doi.org/10.1097/00005110-199901000-00005] [PMID: 9921146]

[32] A.L. Fort, and L. Voltero, "Factors affecting the performance of maternal health care providers in Armenia", *Hum. Resour. Health,* vol. 2, no. 1, p. 8, 2004. [http://dx.doi.org/10.1186/1478-4491-2-8] [PMID: 15212695]

[33] J.K. Swift, and R.P. Greenberg, "Premature discontinuation in adult psychotherapy: A meta-analysis", *J. Consult. Clin. Psychol.,* vol. 80, no. 4, pp. 547-559, 2012. [http://dx.doi.org/10.1037/a0028226] [PMID: 22506792]

[34] A.F. Corning, and E. V. Malofeeva, "The application of survival analysis to the study of psychotherapy termination", *J. Counsel. Psychol.,* vol. 51, no. 3, pp. 354-367, 2004.

[35] B.F. Reis, and L.G. Brown, "Reducing psychotherapy dropouts: Maximizing perspective convergence in the psychotherapy dyad", *Psychotherapy,* vol. 36, no. 2, pp. 123-136, 1999. [http://dx.doi.org/10.1037/h0087822]

[36] M.S. Barrett, W.J. Chua, P. Crits-Christoph, M.B. Gibbons, D. Thompson, and D. Thompson, "Early withdrawal from mental health treatment: Implications for psychotherapy practice", *Psychotherapy,* vol. 45, no. 2, pp. 247-267, 2008.

[http://dx.doi.org/10.1037/0033-3204.45.2.247] [PMID: 19838318]

[37] M.S. Barrett, W. Chua, P. Crits-Christoph, M.B. Gibbons, D. Casiano, and D. Thompson, ""Early withdrawal from mental health treatment: Implications for psychotherapy practice: Correction to Barrett *et al.* (2008)"", *Psychotherapy,* vol. 46, no. 2, p. 248, 2009. [http://dx.doi.org/10.1037/a0016184] [PMID: 22122621]

[38] G.T. Hatchett, and H.L. Park, "Comparison of four operational definitions of premature termination", *Psychotherapy,* vol. 40, no. 3, pp. 226-231, 2003. [http://dx.doi.org/10.1037/0033-3204.40.3.226]

[39] J.K. Swift, and R.P. Greenberg, "Premature discontinuation in adult psychotherapy: A meta-analysis", *J. Consult. Clin. Psychol.,* vol. 80, no. 4, pp. 547-559, 2012. [http://dx.doi.org/10.1037/a0028226] [PMID: 22506792]

[40] N.B. Hansen, M.J. Lambert, and E.M. Forman, "The psychotherapy dose-response effect and its implications for treatment delivery services", *Clin. Psychol. Sci. Pract.,* vol. 9, no. 3, pp. 329-343, 2002. [http://dx.doi.org/10.1093/clipsy.9.3.329]

[41] K.I. Howard, S.M. Kopta, M.S. Krause, and D.E. Orlinsky, "The dose–effect relationship in psychotherapy", *Am. Psychol.,* vol. 41, no. 2, pp. 159-164, 1986. [http://dx.doi.org/10.1037/0003-066X.41.2.159] [PMID: 3516036]

[42] N.B. Hansen, M.J. Lambert, and E.M. Forman, "The psychotherapy dose-response effect and its implications for treatment delivery services", *Clin. Psychol. Sci. Pract.,* vol. 9, no. 3, pp. 329-343, 2002. [http://dx.doi.org/10.1093/clipsy.9.3.329]

[43] E.M. Anderson, and M.J. Lambert, "A survival analysis of clinically significant change in outpatient psychotherapy", *J. Clin. Psychol.,* vol. 57, no. 7, pp. 875-888, 2001. [http://dx.doi.org/10.1002/jclp.1056] [PMID: 11406801]

[44] N.D. Volkow, D. Tomasi, G.J. Wang, F. Telang, J.S. Fowler, J. Logan, H. Benveniste, R. Kim, P.K. Thanos, and S. Ferré, "Evidence that sleep deprivation downregulates dopamine D2R in ventral striatum in the human brain", *J. Neurosci.,* vol. 32, no. 19, pp. 6711-6717, 2012. [http://dx.doi.org/10.1523/JNEUROSCI.0045-12.2012] [PMID: 22573693]

[45] The Hindu, " Mini Tejaswini(2021) "Attrition in IT sector to cross 1 million this year", The Hindu, 27th September",

[46] The Hindu Businessline Infosys CEO sees moderation in attrition over coming quarters by Press Trust of India, 2022.

[47] "Business Today, Why Infosys", *TCS, Wipro, and other Indian IT giants are facing record-high attrition rates,* 2022.

CHAPTER 10

Deep Learning Techniques for Predicting Changes in the Ecosystem

Kruthi[1], **Anugraha Anil kumar**[1], **Aromal A. J.**[1] and **Chaya Ravindra**[1,*]

[1] *School of ECE, REVA University, Bengaluru, India*

Abstract: In the modern age, we depend on technology for our daily needs, from groceries to booking tickets for rides. The technology supports us by understanding our requirements. This is done by using Machine Learning. Machine Learning deals with understanding human behavior and providing suggestions for our requirements. The changes in the ecosystem affect the living creatures who depend on the ecosystem. One of the subsets of Machine Learning that play a vital role in saving the lives of living creatures is Deep Learning. Deep Learning is a representational learning of artificial neural networks. Deep Learning keeps on improving so that it can imitate human intelligence more accurately. Artificial Neural Networks is another subset of Machine Learning and helps in the growth of Deep Learning. There are different classes of artificial neural networks, two of the important classes are the Convolutional Neural Network (CNN) and the Recurrent Neural Network (RNN). The patterns in images are recognized by CNN. So, CNN majorly deals with image recognition and processing. RNN helps recognize sequential data and uses this pattern of sequential data to predict the likely scenarios in the ecosystem. The model, which uses the algorithm of RNN and CNN, should be trained and tested with the data for better efficiency.

Keywords: Artificial neural network, Convolutional neural network, Deep learning, Earthquake, P-waves, Seismic waves, S-waves.

INTRODUCTION

An ecosystem is a place where living organisms, like plants, animals and human beings, and other factors, such as weather and landscape, together form a life. Humans make up most part of the ecosystem, resulting in a negative impact on the ecosystem. Due to this, it also affects other living beings as well. Technology is expanding day by day, and we can use it to save the ecosystem. Machine Learning

* **Corresponding author Chaya Ravindra:** School of ECE, REVA University, Bengaluru, India;
E-mail: Universitychaya@reva.edu.in

Santosh Kumar Das, Soumi Majumder and Nilanjan Dey (Eds.)

and Deep Learning are technologies that can be used to save the ecosystem. Using these technologies, we can measure the longitude, latitude and magnitude through which we can predict the changes in the ecosystem at the earliest.

SEISMIC WAVES

A seismic wave is a vibration that travels through the surface of the Earth's surface and is generated by the motion of the Earth's tectonic plates. It may also be caused by volcanic activities, explosions or other energetic sources. Seismic waves are produced by an earthquake and move through or close to the Earth's surface. Seismic waves are divided into four different categories [1].

P waves, also called primary waves, are the earliest waves that seismographs can detect. The ground is pushed and pulled in the wave's direction by compressional or longitudinal waves. Typically, they do very little harm [2].

S waves, also known as secondary waves, arrive after P waves because of their slower rate of motion. Even though they move in the same direction as the wave, they shake the earth in an opposite direction. S waves have a higher amplitude than P waves and cause the ground's surface to move both vertically and horizontally, making them more dangerous.

Surface waves move slowly and arrive last. They only move across the surface of the Earth. Two types of surface waves are Love waves and Rayleigh waves. Horizontal movement *i.e.*, backward and forward movement, is produced by Love waves. Vertical and horizontal ground movement waves are produced by Rayleigh waves [9]. These waves can be the most harmful because when they pass, they elevate and lower the land as they move along. Seismographs measure the time required by seismic waves to penetrate through different layers of the Earth.

Waves can be reflected and refracted when they pass through materials with various densities and stiffness. Seismologists can determine the type of material the waves are traveling through based on the different ways the waves behave in various materials.

The findings can help us identify fault planes, as well as the pressures and strains operating on them, and provide us a glimpse of the Earth's core structure. In the search for oil and gas, the same wave behavior can also be applied on a smaller scale by monitoring waves produced by explosions or ground vibrators [8].

As we know, seismic waves are one component of earthquakes, and seismologic methods are used for studies on earthquakes [10]. Seismograph locations are required for dependable earthquake hypocenters and focal mechanisms. Resources have found that one focal depth of epicenter would give accurate hypocentral depth control. To determine source parameters, waveform modeling methods are used for earthquakes of magnitude 4.5.

The earthquake structure can be determined using natural sources such as local, regional, and teleseismic sources. Using forward and inverse models, body and surface waves are generated, and a one-dimensional structure of the model can be obtained. Seismic amplitudes and waveforms are considered for forward modeling, and both forward and inverse modeling are used during the travel time of body waves [20].

The Earth's structure is divided into two parts: seismic refraction or wide-angle reflection and seismic reflection studies. The refraction profiles are explained using a two-dimensional model and forward modeling of travel times and amplitude. Refraction methods provide outstanding estimations of seismic velocities while having a low resolution of structure [10].

For seismic refraction, inverse modeling is used using one, two and three dimensional, which also includes tomography. In the past decade, many geographical insights and the highest resolution of crustal structure have beendetermined by seismic reflection.

Earthquake research analysis requires parameters such as permanent regional seismic networks, temporary dense networks of portable seismographs, and worldwide seismic observatories. Permanent regional networks are employed in areas with a high level of seismic activity or potential seismic hazard.

Due to signal clipping, each seismograph contains a single vertical-component 1-Hz geophone and high-gain analog telemetry circuitry.

The temporary network of movable seismographs may be instated after a major earthquake at the epicenter, where these networks have generally 10 to 20 seismographs spaced 5 to 10 km apart, similar to perpendicular-elcment smoked-drum or pen-and-ink seismographs. But nowadays, the networks have three-element seismographs and accelerometers, which give further information for the determination of parameters like magnitude, locations, *etc* [30].

To attain information on hypocentral depth and natural falling on earthquakes greater than the magnitude of 4.45 or more, studies on broadband waveforms are carried out.

Seismic waves produced by earthquakes can cause damage to the Ecosystem. Ground shaking is the ground vibrating because of an earthquake. Earthquakes cause other effects, such as liquefaction and landslides, which are brought on by the shaking. The seismic waves that penetrate underneath buildings, houses, roads, and other structures cause most of the earthquake damage.

Surface fracture is the main threat associated with earthquakes. This can be the result of vertical or horizontal movement on either side of an interrupted fault. Buildings, houses, roads, railroads, and pipelines can be severely damaged by ground shifting, affecting large geographic areas. Bridges, buildings, pipelines, and roads can become unstable from liquefaction, subsidence, collapse, or dissolution [23].

The result of an earthquake that happens beneath the surface of the ocean is termed a tsunami, which is a chain of incredibly lengthy waves. Large tsunamis that rise from the ocean floor are dangerous to buildings, property, infrastructure, and human health. Beyond the coastline, tsunami destruction can be experienced for a long duration [23, 27 - 29].

According to seismic damage data, earthquake-caused fires are the second most common hazard. Due to the earth's tremors, electrical and gas lines become loose, resulting in earthquake fires. Gas is released when gas lines burst, and a spark ignites a firestorm.

LITERATURE SURVEY

Saad *et al.* have focused on classifying the parameters of earthquakes. The authors have used a deep learning approach. They have highlighted the necessity of classifying the parameters. The authors have used CNN and LSTM. They have proposed an algorithm for analyzing the magnitude using CNN, analyzing the dataset and precising the location using LSTM [1].

Carratu *et al.* have focused on the classification of the data set required for predicting the disaster. They have highlighted the installation of seismometers in different places. These seismometers help in detecting calamities early, which helps in warning the people near the affected area so that it can help in saving lives. Here, the authors have used LSTM and CNN. They have also explained how to process the seismic waves and proposed architecture [2].

Arunadevi *et al.* have focused on predicting the risk of earthquakes using Machine Learning. They have talked about the intensity of earthquakes and the damage they can cause. They have also highlighted the case studies on earthquakes, which took place in different locations all around the world. They have spoken on the intensity in each location. Authors have proposed a system design and algorithm [3].

Yoon *et al.*, in their paper that explains seismology, have proposed that data-driven science where breakthroughs often come with advances in observational capabilities. In this paper, we have spoken about seismology, how it helps in various uses and in which forms we can seek data. Data can be in audio, image, or numbers. P-waves and s-waves help in seismology. STA/LTA helps in successfully identifying seismic waves [4].

Scowen *et al.* have proposed that Machine Learning can help in studying the ecosystem. Machine Learning is used for describing data and predictive modeling. The authors speak about the capacity of Machine Learning in the big ecosystem. They have highlighted the best ongoing use of Machine Learning in ecosystem research. Morales-Esteban and Reyes have proposed separate seismic criteria for earthquake prediction in Chile and Iberia, respectively, using mathematical calculations over a time interval of 8-9 days [5].

S.K. Das *et al.* research has proposed an intelligent decision-making scenario that takes into account both cooperative and noncooperative circumstances in both crisp and fuzzy surroundings. Lastly, using the Python programming language, the suggested strategy was simulated and contrasted in terms of payout metrics for outperforming the outcomes [32].

S. K. Das *et al.* have focused on the model of the relationship between earthquake events and parameters, utilizing Bath's law and Omori's law. In Iran, Zamani have suggested a combination of neural networks and mathematical logic to forecast earthquakes [34]. Majumder and Dey's study involved the normalization of seismicity indices and feature extraction, followed by principal component analysis [33].

S.K. Das *et al.*, in their paper, have used payoff matrices. Game theory is applied to describe the nodes' competing strategies. The estimation of parameters relating to uncertainty is done using non-linear geometric programming [35].

A.K. Das *et al.* have concentrated on wireless sensor node safe data transformation. The complexity of the intrusion detection system using sensor nodes comes from the fact that power consumption affects both the network lifetime and the sensor lifetime [36]. By shifting the computational strain of the operation from the sensor nodes to the base station, security and network lifespan can be improved [37].

Another earthquake prediction model in Iran has been proposed by Mirrashid, which incorporates symbolic logic, fuzzy C-means, subtractive clustering, and grid partitioning. This model attempts to predict earthquakes by training various Machine Learning models on seismic and acoustic data obtained from a laboratory micro earthquake simulation [42].

Panakkat and Adeli have introduced an intriguing earthquake prediction method that utilized mathematically derived seismic indicators based on the spatial variation of historical seismic events in Southern California. Their algorithm generated monthly earthquake predictions, and Artificial Neural Networks were used to model the parameters. To establish a sufficient earthquake database, the estimation of all necessary parameters was required. This involved measuring the parameters of seismic events a limited number of times before taking the month into account. Building upon this study, Adeli and Panakkat collaborated to develop earthquake forecasts using the same seismic parameters with the aid of a Probabilistic Neural Network [43].

TYPES OF SEISMIC WAVES

There are two basic types of seismic waves:

P-waves, S-waves.

P-waves and S-waves are sometimes collectively called body waves.

P-WAVE

Primary waves are the fastest waves among various seismic waves, so they are sensed first on the seismograph. P-waves are also known as longitudinal waves because the suspension of the medium either occurs parallel to or in the opposite direction from the direction of the wave's propagation. Because they create compression and rarefaction as they move through a medium, waves are referred to as compressional waves or pressure waves because they cause changes in the medium's pressure. The material stretches (rarefies) and compresses (compresses) because of density changes brought about by P-waves.

These seismic waves are the least destructive and have a reasonably high frequency. These waves are causing the earth's surface to tremble in an upward and downward motion (vertical). They can move through all materials, and the elasticity or shear strength of the medium affects how quickly they move.

Consequently, the order of the primary wave velocity is Solids > Liquids > Gases.

When these waves reach the atmosphere, they transform into sound waves. In earthquakes, the P-wave velocity is between 5 and 8 km/s. With less than 6 km/s in the crust, 13.5 km/s in the lower mantle, and 11 km/s in the inner core, the exact speed varies depending on the area of the Earth's interior.

P waves for Earthquake Cautioning

By detecting the non-destructive primary waves that travel more quickly through the Earth's crust than the destructive secondary and surface waves, advanced earthquake warning prediction is possible [26, 31, 39].

The delay between the arrival of the P-wave and other destructive waves could be up to about 60 to 90 seconds (about 1 and a half minutes) depending on the depth of focus of the earthquake.

S-WAVE

S-waves are called secondary waves because they are the second wave observed by an earthquake seismograph, following the compressional primary wave (P wave), since S-waves travel slowly in solids. S-waves, unlike P waves, cannot penetrate the Earth's molten outer core, creating a shadow zone opposite their origin.

They can nonetheless penetrate below the solid inner core; when a P wave makes an oblique strike on the border between the molten and solid cores, S waves originate and travel in the solid medium [35].

S waves form P waves when they meet at an oblique angle with the boundary and penetrate through the liquid medium. This feature can be used by seismologists to establish a few physical aspects of the Earth's inner core.

After the P-waves, S-waves reach the surface. These waves have a greater frequency than P-waves and a little bit more destructive force [22]. These waves cause the earth's surface to tremble from side to side (horizontal). Fluids (liquids and gases) cannot be penetrated by S-waves because they cannot withstand shear strains. Through the solid portion of the Earth, they move at variable speeds (according to shear strength).

MACHINE LEARNING

We know that technology is a part of our lives and supports us in different ways and different parts of our lives. One of the main reasons for this is Artificial Intelligence and its subsets. One of the subsets of Artificial Intelligence is Machine Learning which is the subset in which the model learns and then performs the task to see the efficiency of the model [6, 38, 40]. There are different ways in which the machine learns to perform tasks, which are supervised, unsupervised learning and reinforcement learning.

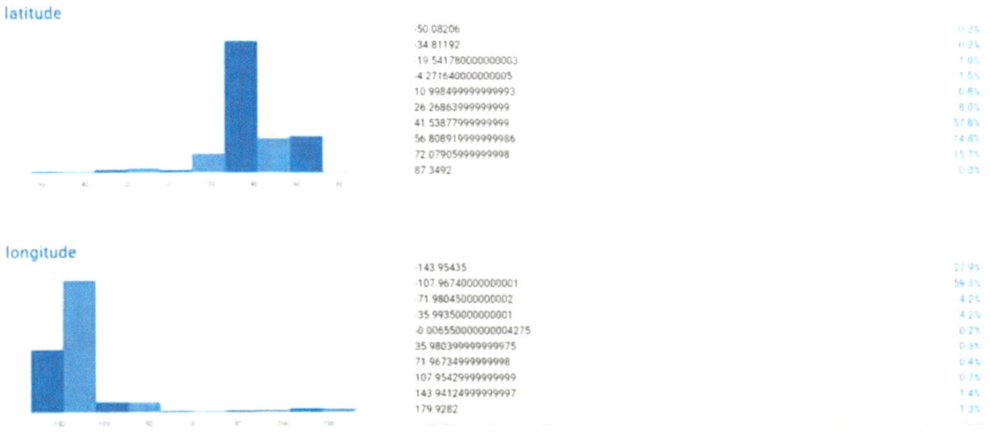

Fig. (1). Latitude and longitude of places where earthquake occurs more.

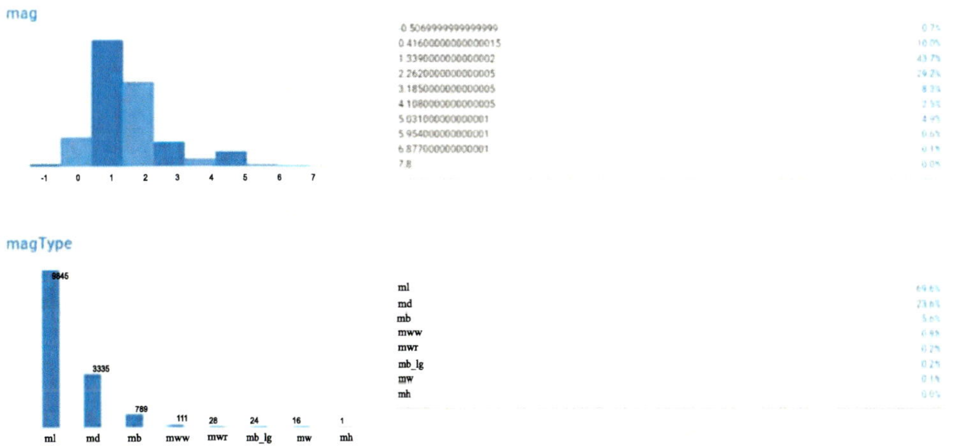

Fig. (2). Number of seismic stations that reported P- and S-arrival times for the earthquake and the magnitude type.

TYPES OF MACHINE LEARNING

Supervised Learning

Supervised learning is where the model is given labeled data during the training. The labeled data act as the supervisor and teach the model to predict the output, and once the model is tested with similar inputs, the model uses its learning from the training data and predicts the output during the testing.

Unsupervised Learning

Unsupervised learning is where the model is trained using unlabeled data, and when the model is tested, it finds some pattern relation in the test data and compares it with the unlabeled data and the output is obtained.

Reinforcement Learning

In reinforcement learning, the model is trained with some behavior or action for a particular environment, and when the model is tested, it gives a positive or negative response according to the actions.

APPLICATIONS OF MACHINE LEARNING

Now, when we take machine learning applications to save our ecosystem, we find that in many parts of our ecosystem, machine learning plays an important role. Some of the applications of machine learning seen in our daily lives are:

Pollution control: IoT devices receive data from different sources. Using machine learning algorithms and predictive models, pollution levels can be estimated and alerts can be given to authorities [7, 41].
Agriculture and food systems: The detection of illness of the crop and the required measures, nourishments to the animals and calculation of yield of crops all can be done using machine learning [20].
Disaster prediction and response: Machine learning is used to predict many natural disasters like earthquakes, floods, tsunamis, *etc* [21]. Since atural disasters can be predicted, the required safety measures can be taken in advance.
Wildlife protection: Many of the animals are now endangered or considered to be extinct.To ensure their safety in their habitat, by using IOT and ML algorithms along with the data of animals like species, age, footprint, *etc.*, we can save many animals from becoming extinct or endangered.

Machine learning also helps in accomplishing one of the key features related to the ecosystem, which is sustainable development [36].

Nowadays, machine learning has an increasing demand and will become one of the critical components in the future, and we see how machine learning plays a vital in our ecosystem [19].

We can also see that early detection of natural disasters can save millions of lives as well as reduce causalities, and this can be done in one way, which is using machine learning [37]. Machine learning has many steps like data cleaning, exploratory data analysis, understanding the data *etc.*, to reach the final output.

These steps are used for the early detection of natural disasters like earthquakes. For the early prediction of earthquakes, the previous number of occurrences of earthquakes along with their magnitude is taken, and using the analysis, the future occurrences of earthquakes and their magnitude can be predicted.

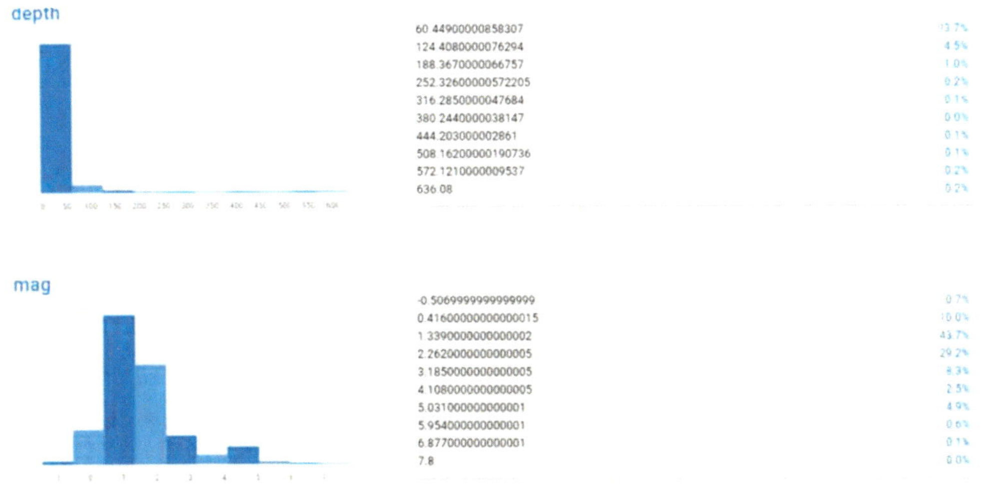

Fig (3). Magnitude and depth of the earthquake.

DEEP LEARNING

As we have seen the role of machine learning in saving the ecosystem, now, let's see the subset of machine learning that is deep learning and how deep learning can play a role in our ecosystem. Deep learning consists of three or more layers of neural networks that simulate the behavior of humans. Deep learning helps the computer to learn complicated concepts by getting trained with simple concepts.

APPLICATIONS OF DEEP LEARNING

When we relate deep learning to the applications for the ecosystem, we find that there are some applications where deep learning is applied, some of which are:

Identification and Classification: In this application, the data is collected from different sources like cameras and recorders, and using deep learning algorithms, the automatic analysis, identification and classification of data is done [25].

Animal Behavior: Using the deep learning algorithms, the insight into collective behavior and social interaction of a species of animal can be observed and identified [24].

Conservation and Management of Ecosystem: As observed, the ecosystem has recently undergone so many changes due to human activities, which cause a lot of trouble to other species that share a part of the ecosystem with humans. These changes can lead to a major loss and depletion of the ecosystem. To avoid the depletion of the ecosystem, what we can do is conserve the ecosystem using the deep learning method. Also, for the smooth working of the ecosystem, it should be managed efficiently using deep learning.

Like machine learning, deep learning is also used in other applications, such as pollution controlling, weather forecasting, *etc.*

As we have seen, both deep learning and machine learning play a great role in the ecosystem. These can be implemented to develop, manage, save and conserve beautiful ecosystems on the Earth.

ARTIFICIAL NEURAL NETWORK

The term "**Artificial Neural Network**" refers to a biologically galvanized sub-field of computing modeled after the brain. It is just like the human brain, which has neurons that are interconnected to one another. Similarly, the Artificial Neural Network has neurons that are interconnected to one another in varied layers of the network. Artificial Neural Network neurons are referred to as nodes.

Artificial neural network plays a crucial role in Machine Learning and Artificial Intelligence technologies. The Artificial Neural Network has three or more layers [11, 12]. The first layer is the input layer, and the final layer is the output layer. The intermediate layers facilitate in analyzing the information. These are put together and referred to as the somatic cell layer/hidden layer. All the inner layers are hidden in the order that they are fashioned by modules that adaptively amend the knowledge obtained from layer to layer through a sequence of transformations [13]. Every layer acts like an associate in the input and output layer that enables the Artificial Neural Network to gain information about a lot of advanced objects.

Artificial Neural Networks are nonlinear applied math models that show a relationship between inputs and outputs. To get a replacement model, different kinds of tools are used, like image recognition, speech recognition, and AI such as in diagnosing. Principally, Artificial Neural Networks are employed for random approximation [14]. Conjointly, they have the ability to obtain sample knowledge instead of the complete datasheet to supply the output result [15]. With Artificial Neural Networks, we are able to enhance existing knowledge analysis techniques as a result of their advanced predictive capabilities.

There are mainly two kinds of Artificial Neural Networks [16]:

- Feedforward Neural Network.
- Feedback Neural Network.

Feedforward Neural Network

The flow of data takes place solely in one direction *i.e.*, the flow of information is from the input layer to the hidden layer and eventually to the output layer. It is employed in supervised learning like classification and image regression [17].

Feedback Neural Network

Feedback loops are a locality of the feedback neural network. This kind of neural network is especially for applications like memory retention in the case of repeated neural networks. These kinds of networks are most relevant to areas where the information is successive or time dependent.

Fig (4). Effects of Earthquakes on the ecosystem are more when compared to other calamities.

CONVENTIONAL NEURAL NETWORK

Conventional Neural Network is one of the types of Artificial Neural Network. It is a network architecture used for facial recognition or image recognition and processing pixel data. The use of Conventional Neural Networks helps in developing applications like self-driving cars [18].

Conventional Neural Networks consist of 3 layers:

- Convolutional layer.
- Pooling layer.
- Full connected layer.

Let's explain the Conventional Neural Network using an example. Let's create a model such that it can identify whether the image is a cat or a dog. For that, the model is trained with dog and cat images, and this is done using thed Conventional Neural Network. Conventional Neural Network captures the features of the image as it collects the data in pixels and then processes it. So, when the image of a dog is detected, it can identify it as a dog [19].

CONCLUSION

The chapter focuses on how deep learning plays a vital role in predicting the changes in the ecosystem, which helps in various ways, such as disaster management, and in the architectural designs of buildings, roads, and others. The deep exploration of different deep learning techniques helps us in exploring new ways of improving the ecosystem, which helps the living beings in the ecosystem to use the ecosystem more efficiently. It has been determined that it is now feasible to use advanced imaging technology to visualize the activity occurring on active fault systems with an unprecedented level of detail. This breakthrough will allow scientists to test existing hypotheses more rigorously and develop new ones. Although it is clear that the new earthquake catalogs contain crucial information about the mechanisms that trigger earthquakes, discovering these mechanisms will require a renewed research effort. In this regard, unsupervised learning techniques may prove to be particularly effective tools for identifying previously unknown patterns and relationships within the data.

REFERENCES

[1] Saad, "Deep learning approach for earthquake parameters classification in earthquake early warning system", *IEEE geoscience and remote sensing letters,* vol. 18, 2021. [http://dx.doi.org/10.1109/LGRS.2020.2998580]

[2] Carratu, *A Deep Learning Approach for the Development of an Early Earthquake Warning System.* IEEE, 2021.

[3] Arunadevi, *et al*, "Risk prediction of earthquakes using machine learning", *Third International Conference on Electronics and Sustainable Communation Systems,* 2022.

[4] Yoon, "Earthquake detection through computationally efficient similarity search", *Sci. Adv.,* 2015. [http://dx.doi.org/10.1126/sciadv.1501057]

[5] M. Scowen, I.N. Athanasiadis, J.M. Bullock, F. Eigenbrod, and S. Willcock, "The current and future uses of machine learning in ecosystem service research", *Sci. Total Environ.,* vol. 799, p. 149263, 2021. [http://dx.doi.org/10.1016/j.scitotenv.2021.149263]

[6] Willock, "'Machine learning for ecosystem services'", *Ecosyst. Serv.,* p. 33, 2018.

[7] Rammer, W., & Seidl, R. (2019). "Harnessing deep learning in ecology: an example predicting bark beetle outbreaks". *Frontiers in plant science*, 10, 451705. [http://dx.doi.org/10.3389/fpls.2019.01327]

[8] Chan, "Detecting the nature of change in an urban environment: A comparison of machine learning algorithms", *Photogrammetric Engineering and Remote Sensing.,* 2021.

[9] Christin, "Applications for deep learning in ecology', canada research chair in polar and boreal ecology', department of biology", In: *University of Moncton, Moncton, NB, Canada 2 Department of Computer Science, University of Moncton, Moncton, NB, Canada*, 2019.

[10] Braspenning, *Artificial neural networks: an introduction to ANN theory and practice* Springer Science & Business Media, 2021.

[11] Dastres, "Artificial neural network systems", *International Journal of Imaging and Robotics,* 2020.

[12] Shrestha, "Review of deep learning algorithms and architectures", *IEEE access.*. [http://dx.doi.org/10.1109/ACCESS.2019.2912200]

[13] Sewak *et al*, *Practical convolutional neural networks: implement advanced deep learning models using Python* Packt Publishing Ltd., 2018.

[14] Chapman, and C, *Fundamentals of seismic wave propagation* Cambridge university press, 2001.

[15] Rikitake and T, "Earthquake Prediction", *Earth Science,* vol. 4, pp. 245-282, 1968.

[16] I.A. Beresnev, "'What we can and cannot learn about earthquake sources from the spectra of seismic waves'", *Bull. Seismol. Soc. Am.,* vol. 91, no. 2, pp. 397-400, 2001. [http://dx.doi.org/10.1785/0120000115]

[17] Li et la, "Machine learning seismic wave discrimination: Application to earthquake early warning", *Geophysical Research Letters,* 2018.

[18] Galkina *et al*, "Machine learning methods for earthquake prediction: A survey", *Proceedings of the Fourth Conference on Software Engineering and Information Management,* vol. 13, p. 25 Saint Petersburg, Russia.

[19] Beroza, "Machine learning and earthquake forecasting—next steps", *Nature communications.,* vol. 12, no. 1, p. 4761, 2021. [http://dx.doi.org/10.1038/s41467-021-24952-6]

[20] Choudhury, *Effects of Earthquake in the surrounding environment-An overview,* 2016.

[21] M.H.A. Banna, K.A. Taher, M.S. Kaiser, M. Mahmud, M.S. Rahman, A.S.M.S. Hosen, and G.H. Cho, "Application of artificial intelligence in predicting earthquakes: State-of-the-art and future challenges", *IEEE Access,* vol. 8, pp. 192880-192923, 2020. [http://dx.doi.org/10.1109/ACCESS.2020.3029859]

[22] Wen and Ding, "Integrated natural disasters risk Management in tourism destination—A case study of 5.12 WenChuan Earthquake", *2010 IEEE International Conference on Management of Innovation & Technology,* 2010.

[23] "Zhang 'Deep learning for seismic lithology prediction,'", *Geophys. J. Int.,* vol. 215, no. 2, pp. 1368-1387, 2018.

[24] Kuyuk and Susumu, "Real-time classification of earthquake using deep learning", *Procedia Comput. Sci.,* vol. 140, 2018.

[25] Bonavolontà, "An innovative air quality monitoring system based on drone and IoT enabling technologies", *2019 IEEE International Workshop on Metrology for Agriculture and Forestry (MetroAgriFor),* IEEE, 2019.

[26] Lior, "P-wave attenuation with implications for earthquake early warning", *Bullettin of the Seismological Society of America,* vol. 106, 2021.

[27] P. Bhargava, "Earthquake prediction using deep neural networks'", *8th International Conference on Advanced Computing and Communication Systems,* 2022.

[28] W.D. Mooney, "Seismic methods for determining earthquake source parameters and lithospheric structure", *GSA Memoirs,* 1989. [http://dx.doi.org/10.1130/MEM172-p11]

[29] Salunkhe, "Prediction of life of piercing punches using artificial neural network and adaptive neuro fuzzy inference systems", *ResearchGate, International Journal of Materials Engineering Innovation,* 2019.
[http://dx.doi.org/10.1504/IJMATEI.2019.097892]

[30] M. Antkowiak, *Artificial neural networks vs. support vector machines for skin diseases recognition,* vol. 21, ResearchGate, pp. 8-2852, 2021.

[31] M. Mondol, "Analysis and prediction of earthquakes using different machine learning techniques", *University of Twente Student Theses,* 2021.

[32] Das and Dey, *Constraint Decision-Making Systems in Engineering.* IGI Global, 2023.

[33] Majumder and Dey, *AI-empowered knowledge management.* Springer, 2022.

[34] S. K. Das, *Smart design and its applications: Challenges and techniques,* pp. 1-6, 2021.
[http://dx.doi.org/10.1007/978-981-33-6195-9_1]

[35] S.K. Das, N. Dey, R. González Crespo, and E. Herrera-Viedma, "A non-linear multi-objective technique for hybrid peer-to-peer communication", *Inf. Sci.,* vol. 629, pp. 413-439, 2023.
[http://dx.doi.org/10.1016/j.ins.2023.01.117]

[36] A.K. Daas, "An optimal data communication system in wireless sensor network using artificial neural network", In: *Constraint Decision-Making Systems in Engineering* IGI Global, 2023.
[http://dx.doi.org/10.4018/978-1-6684-7343-6.ch007]

[37] N. Dey, *Data-Driven Approach for Bio-medical and Healthcare.* Springer Nature, 2022.

[38] Manaswi, "CNN in Keras. Deep learning with applications using python: chatbots and face, object, and speech recognition with tensorflow and keras", *press,* 2018.

[39] "Honda, and H 'Earthquake mechanism and seismic waves.'", *J. Phys. Earth,* vol. 10, p. 2, 2021.

[40] Shivalingagowda, "Wireless sensor network routing protocols using machine learning", *Architectural Wireless Networks Solutions and Security Issues,* 2021.

[41] Tariq, "A real-time early warning seismic event detection algorithm using smart geo-spatial bi-axial inclinometer nodes for industry 4.0 applications", *Appl. Sci.,* vol. 9, no. 18, p. 3650, 2019.

[42] "Earthquake magnitude prediction by adaptive neuro-fuzzy inference system (ANFIS) based on fuzzy C-means algorithm", *Natural hazards,* 2014.

[43] H. Adeli, and A. Panakkat, "A probabilistic neural network for earthquake magnitude prediction", *Neural Netw.,* vol. 22, no. 7, pp. 1018-1024, 2009.
[http://dx.doi.org/10.1016/j.neunet.2009.05.003] [PMID: 19502005]

CHAPTER 11

A Detailed Analysis of Issues with Solid-State Devices

Shweta[1,*] and **Hifzan Ahmad**[2]

[1] *Madan Mohan Malaviya University of Technology, Gorakhpur, Uttar Pradesh, India*

[2] *Dr. A.P.J Abdul Kalam Technical University, Lucknow, India*

Abstract: Non-volatile memory technologies, such as NAND flash memory, have improved storage system performance, reliability, durability, and cost. Due to their speed and density, solid-state devices (SSDs) that are based on flash memory are being used as workstations, desktops, and laptops. Despite offering superior performance, stress resistance, and energy economy as compared to mechanical hard drives, NAND flash memory has unique features and operating limits and cannot be employed as a perfect block device. The design of SSDs has developed over time to make use of the benefits offered by flash memory while, at the same time, hiding their drawbacks. SSD concurrency techniques make use of the available parallelism of flash memories. This chapter thoroughly examines SSD subjects, ranging from the physical features of a flash memory cell to the design of SSDs themselves. The subjects pertaining to the flash translation layer (FTL) are described within the context of interconnected system-level operations. These operations include garbage collection, wear-leveling, address mapping and bad block management. This chapter also provides a review of the most current SSD-related studies.

Keywords: Address translation, Bad block management, Buffer management, Flash memory, Flash translation layer, Garbage collection, Wear leveling.

INTRODUCTION

Recently, flash memory has become a dominant computer and mobile device storage technology. Flash memory-based SSDs are commonly employed as secondary memory in current computers owing to their high-performance hard disk drives (HDDs). HDD performance has stalled owing to slow magnetic platters and actuator arms. SSDs lack complicated mechanical elements, leading to decreased latency and failure rate [1].

SSDs have better bandwidth, random I/O performance, power consumption, stress

* **Corresponding author Shweta:** Madan Mohan Malaviya University of Technology, Gorakhpur, Uttar Pradesh, India; E-mail: shweta20989@gmail.com

tolerance, and system dependability than HDDs [2, 3]. Storage TeK created RAM-based SSDs in 1978. Western Digital launched SSDs in 1989. Early SSDs used the high-performance NOR flash memory structure. M-System created SSDs using NAND flash memory in 1995, but their high cost restricted their deployment (*e.g.*, in industrial or military environments). By 2004, NAND flash memory prices plummeted, and SSDs replaced magnetic storage technologies. Flash memory is utilized in smartphones, tablets, computers, *etc.* Due to their enormous capacity and bandwidth, they are employed in PCs and servers [4]. Despite SSDs' ubiquity and benefits, they have flaws. Due to its physical properties, a flash memory cell must be erased before being programmed. Second, numerous program-and-erase cycles destroy a flash memory cell's oxide layer. This makes flash memory cells unreliable, decreasing their lifespan.

There are two variants of flash memory. Both are known as flash memory, however one is known as NOR flash memory and the second as NAND flash memory. The NOR and NAND flash memory were developed independently by different businesses at separate periods. NOR flash memory is often used to store program code and immediately execute program instructions. This is possible because of the memory's low random-access latency and execute-in-place capability. In lieu of magnetic discs, many consumer devices now make use of NAND memory as an additional storage option. This has been made feasible by the large capacity and low cost of NAND flash memory.

Erase-before-write architecture is one of the fundamental hardware features of flash memory [5]. As a result, FTL system software has been implemented [6]. A table that maps logical addresses to physical addresses is at the heart of an FTL [7]. The mapping table is updated as a result of the newly modified logical/physical address mapping. By doing this, one block is shielded from being overwritten and wiped. There are two main factors to take into account when using the FTL technique in embedded applications: storage performance and RAM capacity constraints. Additionally, embedded programs might benefit from the RAM capacity needed to store the mapping data. Therefore, if an FTL method needs a lot of RAM memory, the product cost will grow. The three different sorts of operations available on flash memory are read, program(write), and erase. The only method to change a bit from 0 to 1 is to erase the block containing the page, which sets all of the block's bits to 1 [8]. Flash has greater performance for random data accesses than HDD since it is built on semiconductor chips that are smaller, consume low energy, and have a lower power need. Additionally, SSDs are heat- and shock-resistant and do not have any moving components or mechanical wear.

However, the workload characteristics have severely affected the relative performance of SSD versus HDD [9, 10]. For certain writing-intensive scientific tasks, for instance, it has been demonstrated that SSDs may only offer a little advantage over HDDs. Additionally, SSD cannot totally replace HDD due to its higher price [11]. Thus, the system-level research issues of flash memory include lifespan extension by reducing the amount of write/erase operations, wear-leveling, handling bad blocks, memory performance improvements through retention relaxation, and hybrid memory design. Fig. (**1**) represents SSD architecture consisting of a buffer manager, RAM and host interface, flash controller and processor. The processor manages the majority of functions, including the flow of requests/responses to mapping. The host interface acts as a physical interface between SSD and the host.

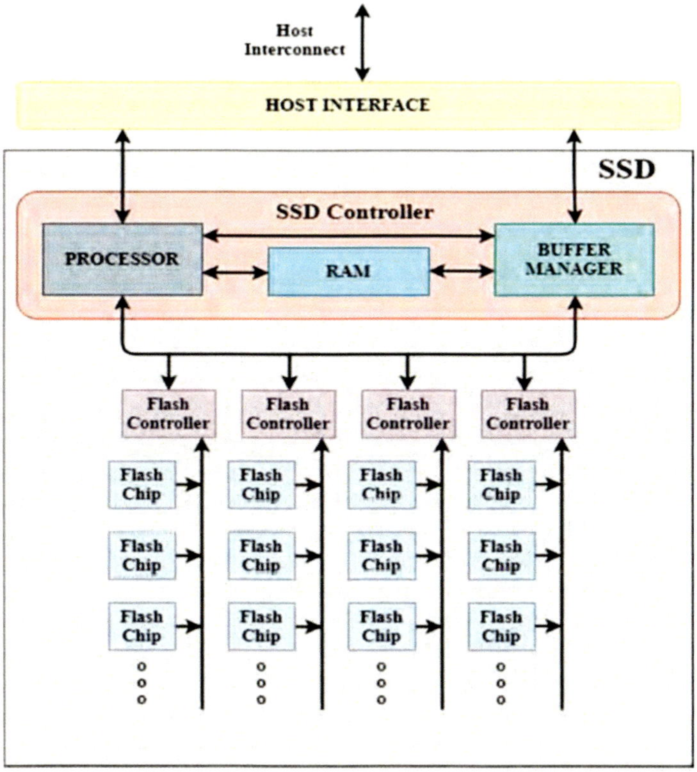

Fig. (1). Architecture of SSD.

SSD designs utilize flash memory's benefits and disguise its drawbacks. To optimize speed, longevity, and reliability, SSDs use FTL.

In comparison to magnetic discs, NAND flash memory exhibits the following five distinguishing characteristics [12]:

No seek time: NAND flash memory is a purely electronic device that does not have any mechanical components, which results in a significant reduction in the amount of time spent seeking data.

The asymmetrical nature of the I/O performances: While the I/O performance of a magnetic disc is virtually the same for write and read operations, the I/O performance of memory varies depending on whether it is used to read, write, or delete data.

Not in place update: To update any data, the block should be first erased. When several erase operations are performed on the same block, the system performance of NAND flash memory suffers as a result. This is because the number of times an erase operation may be performed on a block is limited, and each erase operation requires a significant amount of energy. As a result, NAND flash memory offers an out-of-place update mechanism, which may invalidate the previous data while simultaneously writing the new data on a page that is available.

The maximum number of erase operations that may be performed on each block: In the event when a block's allotted number of erase operations is exceeded, the block in question will experience an increased amount of write failures.

Uses a little amount of energy: NAND flash memory uses far less power than magnetic disc storage. The emergence of NAND flash memory offers tremendous assistance for constructing environmentally friendly data centers and database management solutions with low overall energy consumption.

FLASH TRANSLATION LAYER (FTL)

FTL acts as an interface between the operating system and flash memory. It tries to hide the hardware complexities of memory so that it can be easily used as secondary storage. The following are some of the functions that an FTL algorithm offers [6]:

Address mapping: The primary purpose of an FTL algorithm is to transform the logical addresses generated into the corresponding physical addresses used by flash memory.

Power-off recovery: Even in the case that a sudden power-off event happens while FTL operations are being performed, it is important that FTL data structures be kept and data consistency be assured.

Wear-leveling: It is a function that should be included in FTL. Its purpose is to ensure that memory blocks are worn down as equally as feasible [19, 20].

The above-mentioned functions of FTL are described briefly in the next section. In Fig. (**2**) the functions of FTL are shown.

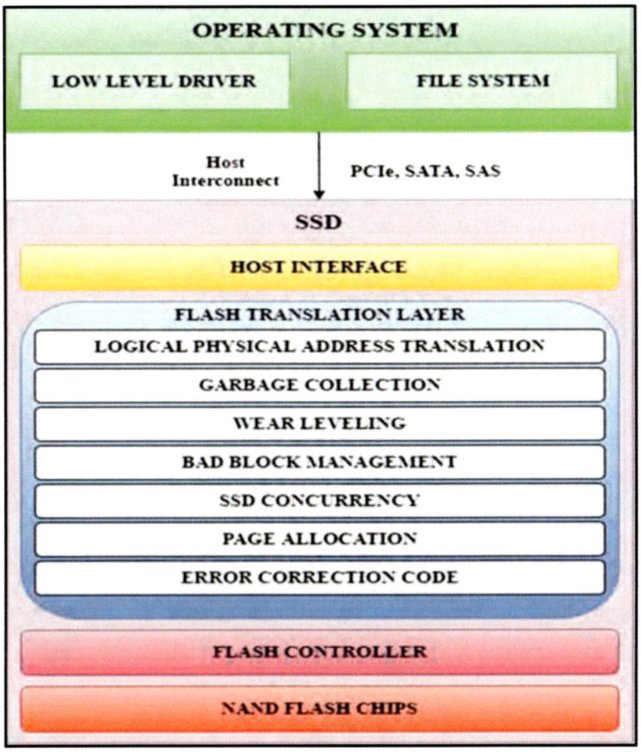

Fig. (2). Functions of flash translation layer.

GARBAGE COLLECTION

Garbage collection is initiated when the flash memory's free space falls below a certain threshold [20]. The procedure of garbage collection typically consists of three phases. It initially chooses the victim blocks, after which it finds the valid pages in it. Finally, it copies the valid pages to some free space, and the victim block is erased for further use [21].

Fig. (**3**) depicts the importance of the efficient garbage collection procedure. As shown in scenario 1, the garbage collector selects BLOCK 1 and 2 as victims in step 1. In step 2, BLOCK 2 consists of 3 valid pages, which are copied to some free blocks *i.e.*, BLOCK 7. In step 3, BLOCK 1 and 2 are erased and ready to be used again. So, there are three write and two erase operations performed by the garbage collector.

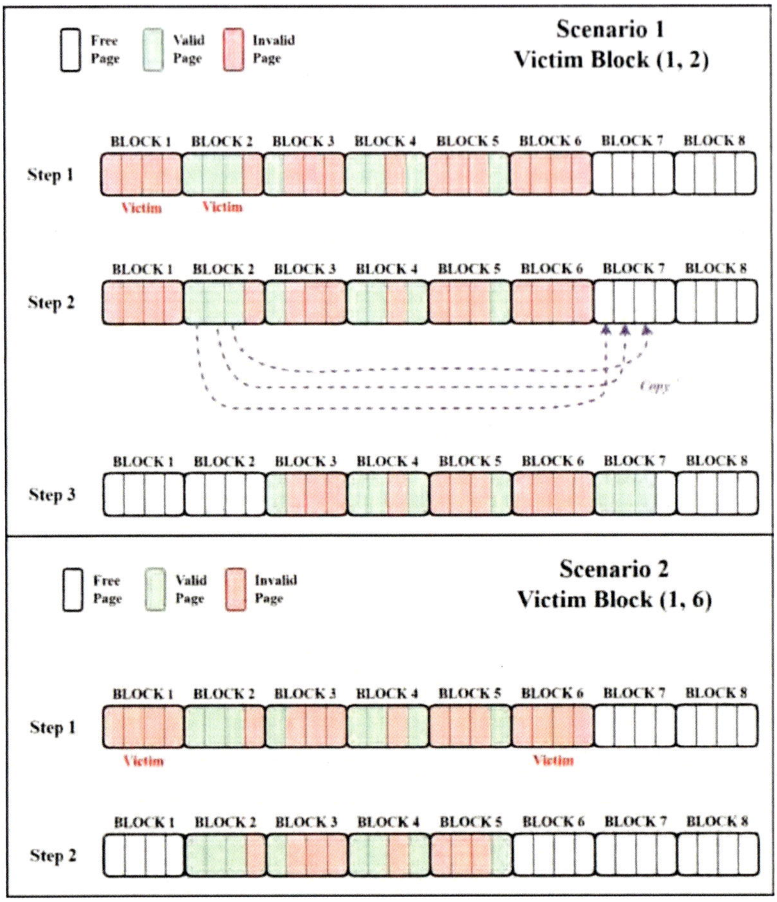

Fig. (3). Need for efficient garbage collection procedure.

In scenario 2, for step 1, the garbage collector selects BLOCK1 and 6 as victims. In the next step, as there are no valid pages in the selected blocks, so there is no write operation. Only two erase operations on BLOCK 1 and 6 are performed. Therefore, by comparing both the scenarios, it is clear that scenario 2 is far better than scenario 1. There are various garbage collection approaches described below:

Greedy Garbage

The garbage collection algorithm has been investigated to reduce cleaning costs and improve wear-leveling to prolong its life. A greedy algorithm (GR) selects a block having the least valid pages as a target/victim block [22]. During this process, GR reduces the cost of migrating valid pages from the victim block to some other free space and the victim block is erased for further use. It performs poorly in the context of wear leveling for significant locality of references since it does not include erase count when selecting a victim block. This lowers the cost of copying legitimate pages in the target block in order to free up space during garbage collection. The greedy approach does not take wear leveling into consideration. The GR technique levels outlast random memory usage but not those with good spatial locality.

Cost Benefit (CB)

CB evaluates cost-benefit values for each and every block and then chooses the block having the maximum value as a victim [23]. It chooses the block having the highest benefit value, where x: latest data invalidation time and u: % of valid pages.

$$Cost\ benefit = \frac{x(1-u)}{2u}.$$

The CB method might enhance wear leveling in flash-based consumer electronics by considering both the number of valid pages per block and the age of invalid pages in the block. The CB algorithm's wear-leveling performance is insufficient since it ignores block erase counts.

Cost Age Times (CAT)

Chiang *et al.* created the CAT algorithm, which enhances the cost-benefit approach by incorporating each block's erase count [21]. Each time a garbage collection operation is executed, each of them picks just one victim block, and none of them conducts studies on I/O workload prediction. CAT tries to balance garbage collection with wear-leveling in flash memory. The CAT policy clusters hot, cold, and read-only data to save cleaning time. It delivers even wear by choosing segments for cleaning based on use, data age, and erase operations carried out on the segments. The CAT policy reduces erases. It employs a fine-grained data redistribution approach to segregate cold from hot data for lesser erase operations. The process is comparable to cost-benefit analysis but on a page level. The CAT policy includes wear-leveling. CAT determines the target block based on cleaning cost, data, age, and erase operations to accomplish even-leveling [25, 26, 27, 28].

Cost-Age-Time with Age-sort (CATA)

The suggested CATA approach identifies victim blocks by using data block usage information such as the age of the block, block utilization, and its erase count [28]. In order to prevent interfering with conventional I/O operations, the suggested approach also forecasts the forthcoming workload. The CATA uses the predicted workload to determine the number of victim blocks. If CATA indicates that the number of I/O request arrivals for the next garbage collection is high, it selects one or no victim block. Otherwise, it selects two or three victim blocks according to the predicted number of I/O request arrivals.

Fast and Endurant Garbage Collection (FeGC)

A quick and reliable garbage collection method referred to as FeGC was described for use in storage systems [29]. The primary aim of FeGC is to speed up the garbage collection process while also extending the life of memory devices. In order to do this, it first determines which blocks will be victims of GC by taking into account the instance at which each invalid page in a block becomes invalid, after which it redistributes valid pages according to the differences at which they are updated. In addition, FeGC is capable of effectively managing free blocks and allocating them in a manner that is acceptable taking into account the features of both the pages and the blocks. FeGC chooses the garbage collection victim blocks by taking into account the invalidated times of all the invalid pages in a block and then redistributes the valid pages according to their update intervals. Furthermore, FeGC handles free blocks effectively and assigns them correctly based on the features of the pages and blocks. Fig. (**4**) describes the working of the FeGC algorithm.

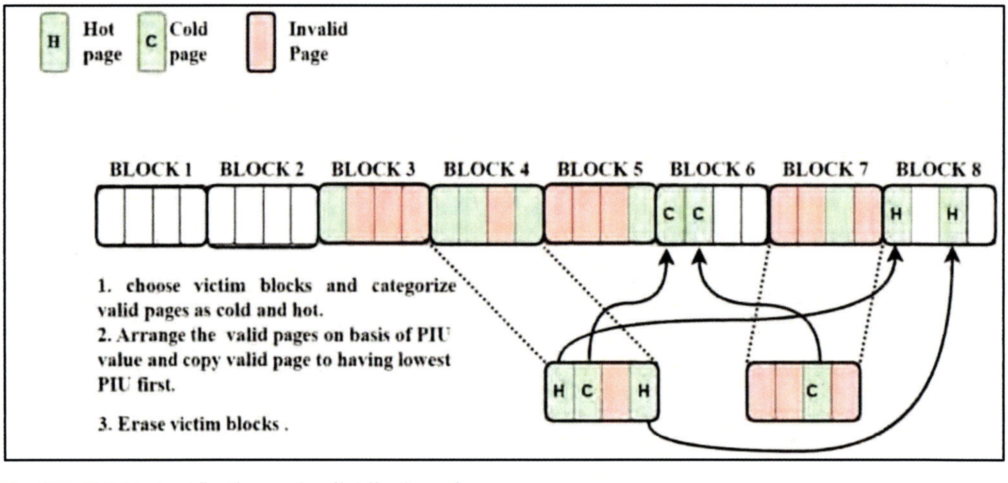

Fig. (4). FeGC classification and redistribution of pages.

/*if flash memory does not have enough space, GC() is invoked*/

GC()

{

If there is no space {

for each block, calculate the cost using formula

$$Value = \sum_{x=1}^{n} Age$$

$Age = T_c - T_i$

Select the block having the highest value of cost as the victim.

}

Swap-Aware Garbage Collection (SAGC)

In order to effectively organize the swap space for portable electronic devices, SAGC is suggested as a new garbage collection strategy [30]. The swapping I/O properties for a swap system based on NAND flash memory are taken into account by the SAGC. Reduced write and erase operations are attempted as part of the suggested trash collection policy, which also takes wear-leveling into account for NAND flash memory persistence. Three characteristics of the SAGC are crucial in NAND flash memory-based swap systems: (1) extended NAND flash memory endurance; (2) efficient garbage collection; and (3) low power consumption. Fig. (**5**) depicts the procedure of victim selection in SAGC.

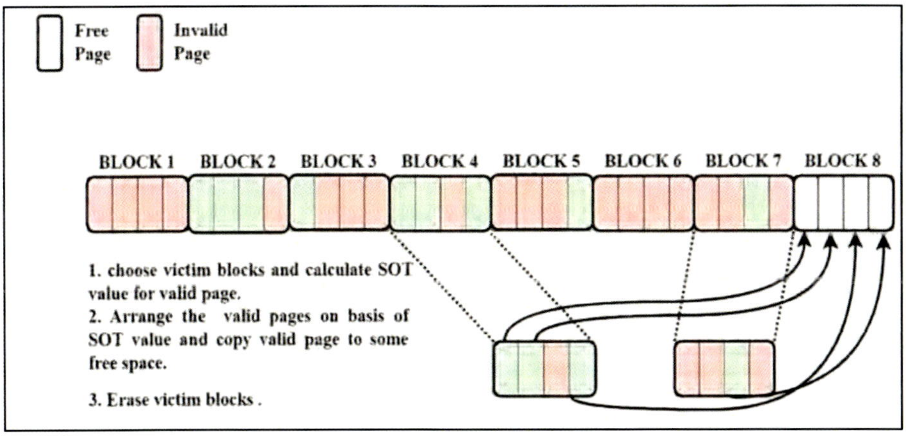

Fig. (5). Swap-aware garbage collection victim selection and page distribution procedure.

File-Aware Garbage Collection (FaGC)

In accordance with the estimated update frequency of the corresponding chunk, the FaGC method replicates a victim block's valid pages to clusters in free blocks [31]. To enhance the longevity of flash memory, it uses a hybrid wcar-lcvcling approach. A scattering factor is established and evaluated to decide when to activate the cleaning approach in order to avoid unnecessary garbage collection.

Efficient and Non-Sensitive File-Aware Garbage Collection (FaGC+)

The FaGC+ [32] algorithm uses a unique cold-hot logical page classification system and update frequency calculation technique. To determine how frequently a logical page is modified, the suggested method uses a new PSN. It chooses the block having the highest CPS as the target block and then divides all of the victim block's eligible pages into four categories *i.e.*, cold, hot, second-hot and second-cold. Fig. (**6**) represents the procedure of data distribution after victim block selection.

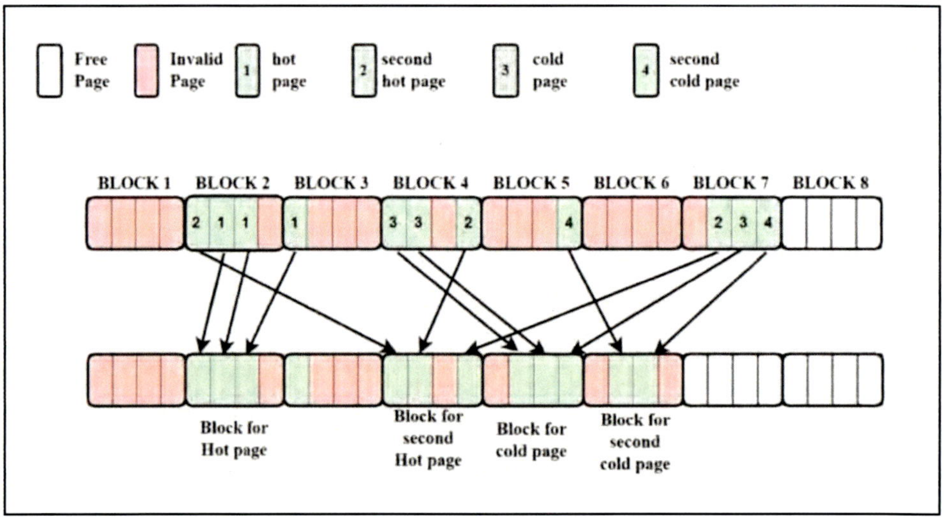

Fig. (6). FaGC+ data redistribution.

A comparison between various cleaning approaches is shown in Table **1**, which describes the techniques used in them.

Table 1. Various garbage collection techniques.

Algorithm	Approach Applied
Greedy [21]	Chooses the block with minimum utilization Where, $$Utilization = \frac{No.of\ Valid\ pages}{Total\ No.of\ pages}$$
Cost Benefit(CB) [22]	Chooses the block having the highest benefit value, where x: latest data invalidation time and u: % of valid pages. $$Cost\ benefit = \frac{x(1-u)}{2u}$$

(Table 1) cont.....

Algorithm	Approach Applied
Cost Age Time(CAT) [19]	Chooses the block having the highest cost Where u: % of valid pages; E: Erase Count; A: age $Cost = \frac{u}{1-u} * \frac{1}{A} * E$
Cost-Age-Time with Age-sort (CATA) [23]	Chooses the block having maximum value Where u= block utilization; Ai: age; E: Erase count $\frac{1-u}{1+u} * A_i * \frac{1}{E}$
Fast and Endurant Garbage Collection (FeGC) [24]	Chooses the block having the highest value $value = \sum_{i=1}^{n} A_i$
Swap-Aware Garbage Collection (SAGC) [25]	Chooses the block with the highest value, $value = \sum_{i=1}^{n} A_i$ Where C_T: Completion time; I_T = Initial time $A_i = C_T - I_T$
File-Aware Garbage Collection (FaGC) [26]	Uses scattering factor
Efficient and Non-sensitive File-Aware Garbage Collection (FaGC+) [27]	Calculates the update frequency for every valid page in the target block.

Address Mapping

On the basis of granularity, address mapping can be further subdivided into page-level mapping or sector mapping, block mapping and hybrid mapping [35].

Sector Mapping

In sector mapping, the logical page is assigned to its physical sector [35]. As a result, the row of the logical-to-physical mapping table is p if the file system defines p logical sectors. In Fig. (7), it is depicted that if the system requests WRITE (8,B), which means to write data B to LSN 8, the mapping is done with the help of a page map table, which associates it to PSN 7 in Block 1. The sector-level mapping approach allows the system to select any valid page for read and write. The page mapping table, on the other hand, must keep track of all the mapping information for the flash memory, which necessitates a size of mapping table $\lfloor \log 2\ ^{N}/8 \rfloor^{*} N$ where N is the number of SSD pages.

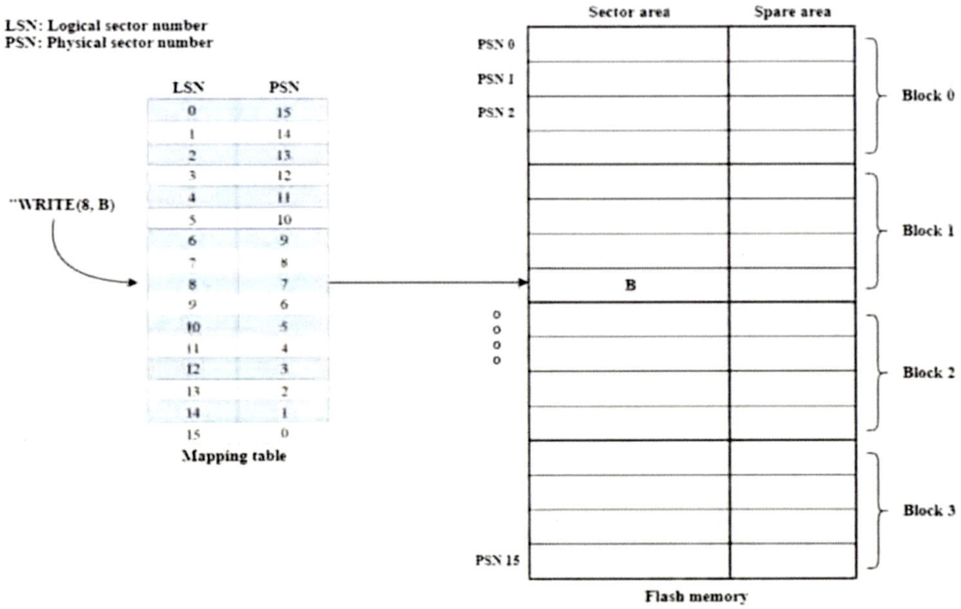

Fig. (7). Sector mapping.

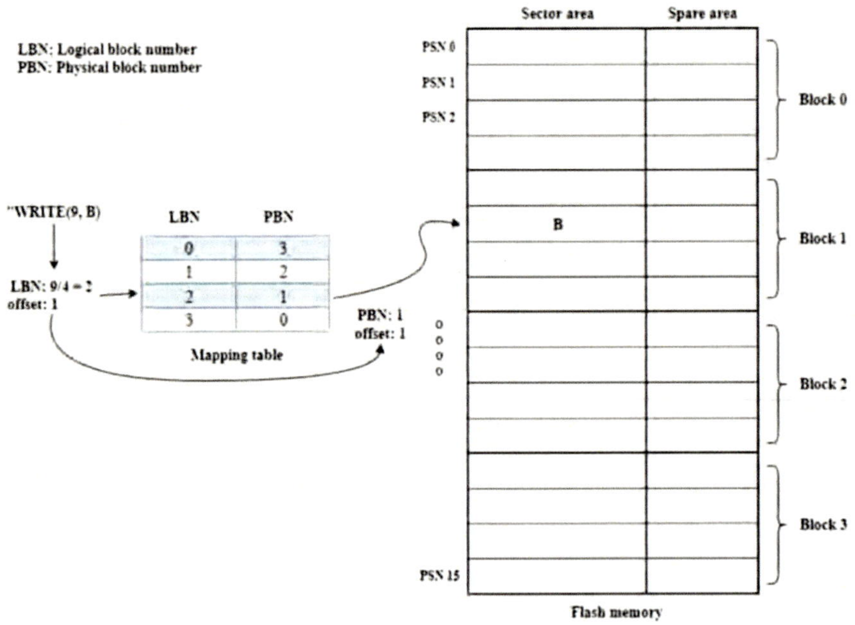

Fig. (8). Block mapping.

Block Mapping

Because the sector mapping process demands a considerable amount of RAM, it is impractical for compact embedded devices [35, 36]. The working of block mapping is shown in Fig. (**8**). Where LBN is converted into offset and PBN. The LBN is divided by the number of pages: for example, in this scenario, it is 4; the quotient of division is a logical block number. Although the mapping table size is reduced by B/P where B is the block size and P is the page size, there are still some disadvantages, such as block mapping requires a fixed offset number in a block. However, it reduces the table size, but free blocks should be assigned with the same logical page every time and updating causes extra cost. Many copy and erase procedures are necessary if the file system gives write orders with similar logical sector numbers, significantly reducing performance. Block-level mapping data should be kept in flash memory when implementing the block mapping technique so that it may be recovered after a power failure.

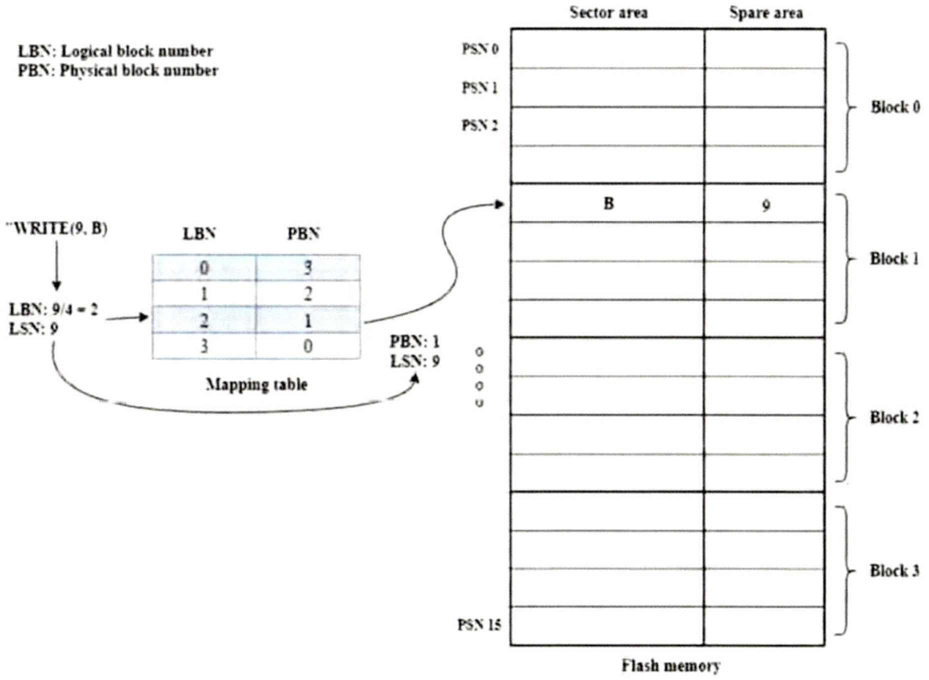

Fig. (9). Hybrid mapping.

Hybrid Mapping

As the name says, hybrid mapping employs block mapping to get the physical block [37]. Then, it utilizes sector mapping to find an empty sector in the physical block. It maintains two blocks *i.e.*, data and log. The data block holds normal data, whereas the log block holds updated requests. The former is organized by block-level mapping, whereas the latter is defined by page-level mapping.

An overview of the hybrid approach is shown in Fig. (**9**). The FTL method determines the logical block number 2(=9/4) for the LSN when the file system gives the command "write(9,A)", then it locates the physical block number 1 from the mapping table. The FTL algorithm assigns an empty sector for the update after getting the physical block number.

The GC overhead for three address mapping techniques is:

Sector << Hybrid << Block

The memory requirement is:

Sector >>> Hybrid > Block

WEAR LEVELING

A flash memory block only withstands a certain number of erase operations (usually 500–3000) before being worn out and damaged because of the memory's worn-out feature. The GC and other flash memory management tools must thus be carefully maintained for an effective wear-leveling effect *i.e.*, to perform load balancing to each block so that it prolongs the lifespan of flash memory, in addition to the performance difficulties [38, 39].

Flash blocks can be erased to some fixed threshold (*i.e.*, the maximum endurable P/E cycles), which motivates the adoption of wear-leveling designs. The wear-leveling design evenly distributes block erases throughout the whole NAND flash memory area to prevent some flash blocks from prematurely wearing out. This increases the persistence of flash devices and extends the device's lifetime. There are now two major categories of wear-leveling strategies.

1. Dynamic wear leveling [40]: This type of algorithm maintains wear leveling by recycling blocks with lower erase counts several times. These techniques do not try to relocate cold data, which could stay in a few blocks for all time. Compared to conventional blocks, these cold data storage blocks deteriorate relatively slowly.

2. Static wear leveling: Unlike dynamic wear-leveling algorithms, static wear-leveling algorithms make an effort to transfer cold data to worn-out blocks to promote a more even distribution of wear [41]. Moving cold data along, nevertheless, not having any request for modification, results in overhead.

The need for both dynamic and static wear-leveling is demonstrated in Fig. (**10**) in order to prevent wearing-out properties of flash blocks. Blocks that are updated frequently are known as hot, whereas blocks that are changed infrequently are known as cold. The cold block's and hot block's respective P/E cycle counts in this illustration are 1k and 5k. Only the hot block will wear out if the dynamic wear-leveling approach is employed solely for 5k P/E cycles since the cold block is been updated.

Evenness-Aware Algorithm

The evenness-aware method is designed to avoid static data persisting at any block for an extended time [43]. As a result, it reduces the max erase count between any two blocks; therefore, the lifetime of flash memory is increased. This algorithm manages a table known as a block erasing table (BET), which keeps track of blocks that have been erased over a certain amount of time. For the purposes of static wear-leveling, the BET is connected to the SW Leveler, based on the BET information, so when SW Leveler is activated, it either resets the BET or selects a block that is yet to be deleted, causing garbage collector to perform its function, and when the block is deleted, the BET table is changed as per the requirement.

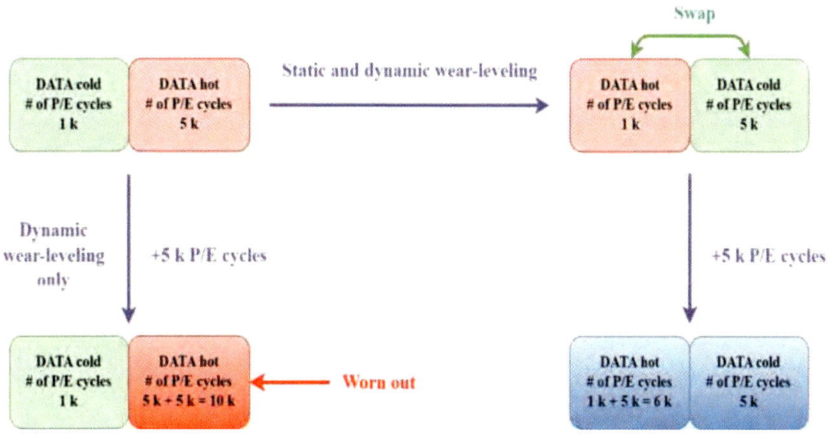

Fig. (10). Static and dynamic wear-leveling.

Dual Pool Algorithm

L.P. Chang's dual pool algorithm maintains two block pools: a hot pool and a cold pool [42]. The cold and hot pools are first given the blocks at random. Blocks that retain hot data are linked with the hot pool, and blocks that hold cold data are connected with the cold pool. When updates are made, the pool affiliations become stable. The contents of the least-worn block in the cold pool are swapped with those of a hot pool block whose contents have been deleted over a specific threshold. The pool associations of blocks take a very long time to stabilize in the algorithm. Before the blocks are successfully linked to the proper pools, there may be several data migrations.

BAD BLOCK MANAGEMENT

Bad blocks are those in which at least one invalid bit is present and whose reliability is uncertain [43]. Bad blocks could already be there when the gadget is supplied, or they might appear later on. Flash blocks may have defects during manufacturing or wear out after a certain amount of P/E cycles. To preserve high dependability, poor block management stays away from these bad blocks.

Detection of Bad Block

Bad blocks that develop during manufacturing are identified during factory testing by subjecting flash devices to extreme conditions as well as *via* exclusive test modes [44]. The makers supply location information to these blocks, which must be read before carrying out any erase operations because it is erasable and cannot be restored once lost. These blocks may be identified by determining whether or not the data at particular points is FFh. Bad blocks formed when program/erase cycles exceed the endurance limit can be discovered by detecting the program error or erase error that occurs during the program or erase operation.

Replacement of Bad Block

Further bad blocks may form during the life of the NAND flash memory. A status register on NAND flash indicates the success of an operation. Further bad blocks are found when efforts to PROGRAM or ERASE result in failures in the status register. The block can be replaced by reprogramming the present data and transferring the remainder of the replaced block to an available valid block since the inability of a PAGE PROGRAM operation does not impact the data in subsequent pages in the same block. There are two common ways to assign fresh blocks and designate problematic blocks:

1. Skip block approach.

2. Reserve block approach.

Skip Block Approach

In the skip block approach, data is put in the next good block that becomes available instead of the bad block addressed if the address received from the host matches any of the addresses recorded in the BET. Bad blocks produced over the lifespan of a flash device will also be placed in the database, thus, when they are addressed, these blocks will be skipped.

Reserve Block Approach (RBA)

RBA substitutes bad blocks by forwarding to a specified free good block [45] . In this method, NAND flash is separated into two regions: the reserved block region and the user addressable block region, as shown in Fig. (**11**). The reserved block region is solely used to replace bad blocks and save the bad block table. It also maintains track of the remapped and generated bad blocks. User-addressable block regions are used to store data supplied by hosts.

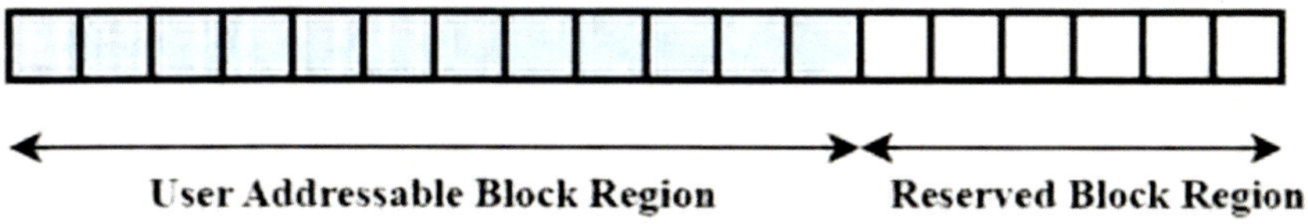

Fig. (11). FaGC+ data redistribution.

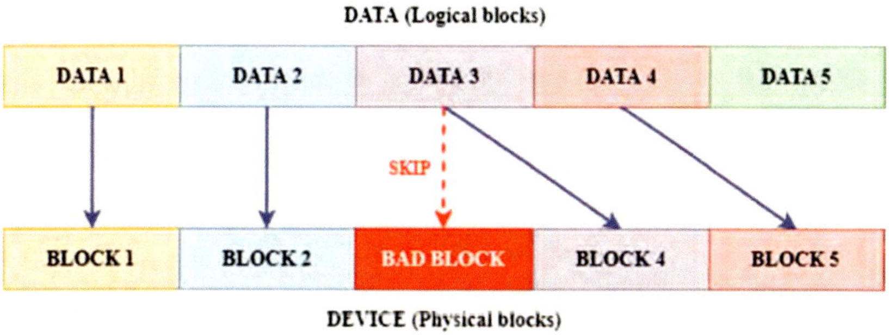

Fig. (12). Skip block approach.

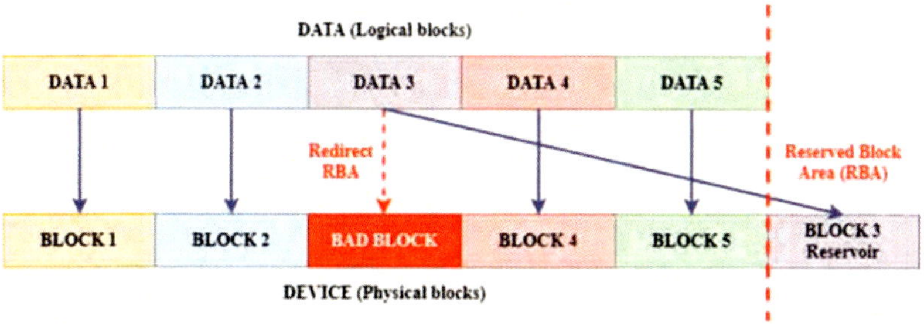

Fig. (13). Reserve block approach.

As depicted in Fig. (**12**) BLOCK 3 is faulty, so it is skipped, and BLOCK 4 is picked instead of the faulty block. When a defective block is detected, the RBA selects a nonfaulty block to write the data. Fig. (**13**) depicts the RBA. As in the previous example, BLOCK 3 is recognized to be a bad block in this case, therefore DATA 3 is written from the reserved block region to a nonfaulty block (referred to as a reservoir block).

CONCLUDING REMARKS

This chapter covered every aspect of SSD difficulties, including all the fundamental FTL features. Additionally, it presented the most recent advancements in research in various areas covered in this chapter for enhancing performance, endurance, and reliability. Our study illustrated the substantial and closely intertwined nature of the research difficulties in SSDs. The performance of forthcoming SSD systems will thus need to be improved by addressing these problems comprehensively.

REFERENCES

[1] R. Micheloni, "Inside solid state drives (SSDs). Edited by Alessia Marelli, and Kam Eshghi", *Heidelberg/Berlin, Germany: Springer.,* 2013.
[http://dx.doi.org/10.1109/TCE.2002.1010143]

[2] N. Agrawal, "Design Tradeoffs for {SSD} Performance", *In 2008 USENIX Annual Technical Conference.,* 2008.
[http://dx.doi.org/10.1109/TCE.2002.1010143]

[3] C. Dirik, "The performance of PC solid-state disks (SSDs) as a function of bandwidth, concurrency, device architecture, and system organization", *ACM SIGARCH Computer Architecture News 37.,* pp. 279-289, 2009.
[http://dx.doi.org/10.1109/TCE.2002.1010143]

[4] U.K. Jeong, "A superblock-based flash translation layer for NAND flash memory", *A superblock-based flash translation layer for NAND flash memory.,* pp. 161-170, 2006.
[http://dx.doi.org/10.1109/TCE.2002.1010143]

[5] Jesung Kim, S.H. Noh, Yookun Cho, S.L. Min, and Y. Cho, "A space-efficient flash translation layer for CompactFlash systems", *IEEE Trans. Consum. Electron.,* vol. 48, no. 2, pp. 366-375, 2002.
[http://dx.doi.org/10.1109/TCE.2002.1010143]

[6] T.S. Chung, D.J. Park, S. Park, D.H. Lee, S.W. Lee, and H.J. Song, "A survey of flash translation layer", *J. Systems Archit.,* vol. 55, no. 5-6, pp. 332-343, 2009.
[http://dx.doi.org/10.1016/j.sysarc.2009.03.005]

[7] A. Gupta, Y. Kim, and B. Urgaonkar, "DFTL: A flash translation layer employing demand-based selective caching of page-level address mappings", *SIGPLAN Not.,* vol. 44, no. 3, pp. 229-240, 2009.
[http://dx.doi.org/10.1145/1508284.1508271]

[8] S. Mittal, and J.S. Vetter, "A survey of software techniques for using non-volatile memories for storage and main memory systems", *IEEE Trans. Parallel Distrib. Syst.,* vol. 27, no. 5, pp. 1537-1550, 2016.
[http://dx.doi.org/10.1109/TPDS.2015.2442980]

[9] F. Chen, D.A. Koufaty, and X. Zhang, "Hystor: Making the best use of solid state drives in high performance storage systems", *Proceedings of the international conference on Supercomputing,* pp. 22-32, 2011.
[http://dx.doi.org/10.1145/1995896.1995902]

[10] Y. Kim, A. Gupta, B. Urgaonkar, P. Berman, and A. Sivasubramaniam, "HybridStore: A cost-efficient, high-performance storage system combining SSDs and HDDs", *In Proceedings of the 2006 international conference on Compilers, architecture and synthesis for embedded systems,* pp. 227-236, 2011.
[http://dx.doi.org/10.1109/MASCOTS.2011.64]

[11] I. Koltsidas, and S.D. Viglas, "Flashing up the storage layer", *Proceedings VLDB Endowment,* vol. 1, no. 1, pp. 514-525, 2008.
[http://dx.doi.org/10.14778/1453856.1453913]

[12] G. Xu, F. Lin, and Y. Xiao, "CLRU: A new page replacement algorithm for NAND flash-based consumer electronics", *IEEE Trans. Consum. Electron.,* vol. 60, no. 1, pp. 38-44, 2014.
[http://dx.doi.org/10.1109/TCE.2014.6780923]

[13] D. Seo, and D. Shin, "Recently-evicted-first buffer replacement policy for flash storage devices", *IEEE Trans. Consum. Electron.,* vol. 54, no. 3, pp. 1228-1235, 2008.
[http://dx.doi.org/10.1109/TCE.2008.4637611]

[14] "Park, Seon-yeong, Dawoon Jung, Jeong-uk Kang, Jin-soo Kim, and Joonwon Lee. "CFLRU: A replacement algorithm for flash memory", *In Proceedings of the 2006 international conference on Compilers, architecture and synthesis for embedded systems,* pp. 234-241, 2006.

[15] H. Jung, H. Shim, S. Park, S. Kang, and J. Cha, "LRU-WSR: Integration of LRU and writes sequence reordering for flash memory", *IEEE Trans. Consum. Electron.,* vol. 54, no. 3, pp. 1215-1223, 2008.
[http://dx.doi.org/10.1109/TCE.2008.4637609]

[16] Z. Li, P. Jin, X. Su, K. Cui, and L. Yue, "CCF-LRU: A new buffer replacement algorithm for flash memory", *IEEE Trans. Consum. Electron.,* vol. 55, no. 3, pp. 1351-1359, 2009.
[http://dx.doi.org/10.1109/TCE.2009.5277999]

[17] Y. Yuan, Y. Shen, W. Li, D. Yu, L. Yan, and Y. Wang, "PR-LRU: A novel buffer replacement algorithm based on the probability of reference for flash memory", *IEEE Access,* vol. 5, pp. 12626-12634, 2017.
[http://dx.doi.org/10.1109/ACCESS.2017.2723758]

[18] S.K. Das, and N. Dey, "Constraint decision-making systems in engineering", In: *IGI Global,* 2023.

[19] S. Majumder, and N. Dey, *AI-empowered knowledge management.* Springer, 2022.
[http://dx.doi.org/10.1007/978-981-19-0316-8]

[20] M.L. Chiang, and R.C. Chang, "Cleaning policies in mobile computers using flash memory", *J. Syst. Softw.,* vol. 48, no. 3, pp. 213-231, 1999.
[http://dx.doi.org/10.1016/S0164-1212(99)00059-X]

[21] M.L. Chiang, P.C.H. Lee, and R-C. Chang, "Using data clustering to improve cleaning performance for flash memory", *Softw. Pract. Exper.,* vol. 29, no. 3, pp. 267-290, 1999.
[http://dx.doi.org/10.1002/(SICI)1097-024X(199903)29:3<267::AID-SPE233>3.0.CO;2-T]

[22] M. Wu, and W. Zwaenepoel, "eNVy", *Oper. Syst. Rev.,* vol. 28, no. 5, pp. 86-97, 1994.
[http://dx.doi.org/10.1145/381792.195506]

[23] A. Kawaguchi, S. Nishioka, and H. Motoda, "A flash-memory based file system", In: *In USENIX,* 1995, pp. 155-164.

[24] S. K. Das, "Smart design and its applications: Challenges and techniques", *Nature-Inspired Computing for Smart Application Design,* pp. 1-6, 2021.

[25] S.K. Das, N. Dey, R. González Crespo, and E. Herrera-Viedma, "A non-linear multi-objective technique for hybrid peer-to-peer communication", *Inf. Sci.,* vol. 629, pp. 413-439, 2023.
[http://dx.doi.org/10.1016/j.ins.2023.01.117]

[26] A.K. Dass, and S.K. Das, "An optimal data communication system in wireless sensor network using artificial neural network", In: *Constraint Decision-Making Systems in Engineering.* IGI Global, 2023,

pp. 120-134.
[http://dx.doi.org/10.4018/978-1-6684-7343-6.ch007]

[27] N. Dey, Ed., *Data-Driven Approach for Bio-medical and Healthcare.* Springer Nature, 2022.

[28] L. Han, Y. Ryu, and K. Yim, "CATA: A garbage collection scheme for flash memory file systems", *International Conference on Ubiquitous Intelligence and Computing,* Springer: Berlin, Heidelberg, pp. 103-112, 2006.
[http://dx.doi.org/10.1007/11833529_11]

[29] O. Kwon, K. Koh, J. Lee, and H. Bahn, "FeGC: An efficient garbage collection scheme for flash memory based storage systems", *J. Syst. Softw.,* vol. 84, no. 9, pp. 1507-1523, 2011.
[http://dx.doi.org/10.1016/j.jss.2011.02.042]

[30] O. Kwon, and K. Koh, "Swap space management technique for portable consumer electronics with NAND flash memory", *IEEE Trans. Consum. Electron.,* vol. 56, no. 3, pp. 1524-1531, 2010.
[http://dx.doi.org/10.1109/TCE.2010.5606292]

[31] H. Yan, and Q. Yao, "An efficient file-aware garbage collection algorithm for NAND flash-based consumer electronics", *IEEE Trans. Consum. Electron.,* vol. 60, no. 4, pp. 623-627, 2014.
[http://dx.doi.org/10.1109/TCE.2014.7027335]

[32] H. Yan, Y. Huang, X. Zhou, and Y. Lei, "An efficient and non-time-sensitive file-aware garbage collection algorithm for NAND flash-based consumer electronics", *IEEE Trans. Consum. Electron.,* vol. 65, no. 1, pp. 73-79, 2019.
[http://dx.doi.org/10.1109/TCE.2018.2885102]

[33] C. Park, W. Cheon, J. Kang, K. Roh, W. Cho, and J.S. Kim, "A reconfigurable FTL (flash translation layer) architecture for NAND flash-based applications", *ACM Trans. Embed. Comput. Syst.,* vol. 7, no. 4, pp. 1-23, 2008.
[http://dx.doi.org/10.1145/1376804.1376806]

[34] J. Huang, A. Badam, M.K. Qureshi, and K. Schwan, "Unified address translation for memory-mapped ssds with flashmap", *In Proceedings of the 42Nd Annual International Symposium on Computer Architecture,* pp. 580-591, 2015.
[http://dx.doi.org/10.1145/2749469.2750420]

[35] A. Tavakkol, P. Mehrvarzy, M. Arjomand, and H. Sarbazi-Azad, "Performance evaluation of dynamic page allocation strategies in SSDs", *ACM Transactions on Modeling and Performance Evaluation of Computing Systems,* vol. 1, no. 2, pp. 1-33, 2016.
[http://dx.doi.org/10.1145/2829974]

[36] G. Wu, and X. He, "Delta-FTL: Improving SSD lifetime *via* exploiting content locality", *Proceedings of the 7th ACM european conference on Computer Systems,* pp. 253-266, 2012.
[http://dx.doi.org/10.1145/2168836.2168862]

[37] L-P. Chang, "On efficient wear leveling for large-scale flash-memory storage systems", *Proceedings of the 2007 ACM symposium on Applied computing,* pp. 1126-1130, 2007.
[http://dx.doi.org/10.1145/1244002.1244248]

[38] X-Y. Hu, R. Haas, and E. Evangelos, "Container marking: Combining data placement, garbage collection and wear levelling for flash", *In 2011 IEEE 19th Annual International Symposium on Modelling, Analysis, and Simulation of Computer and Telecommunication Systems,* pp. 237-247, 2011.
[http://dx.doi.org/10.1109/MASCOTS.2011.50]

[39] M-C. Yang, Y-M. Chang, C-W. Tsao, P-C. Huang, Y-H. Chang, and T-W. Kuo, "Garbage collection and wear leveling for flash memory: Past and future", *2014 International Conference on Smart Computing,* IEEE, pp. 66-73, 2014.
[http://dx.doi.org/10.1109/SMARTCOMP.2014.7043841]

[40] B. Peleato, H. Tabrizi, R. Agarwal, and J. Ferreira, "BER-based wear leveling and bad block management for NAND flash", *2015 IEEE International Conference on Communications (ICC),*

IEEE, pp. 295-300, 2015.
[http://dx.doi.org/10.1109/ICC.2015.7248337]

[41] Y-H. Chang, and L-P. Chang, "Efficient wear leveling in NAND Flash memory", In: *Inside Solid State Drives (SSDs)*. Springer: Dordrecht, 2013, pp. 233-257.
[http://dx.doi.org/10.1007/978-94-007-5146-0_9]

[42] Q. Li, H. Li, and K. Zhang, "A survey of SSD lifecycle prediction", *In 2019 IEEE 10th International Conference on Software Engineering and Service Science,* pp. 195-198, 2019.
[http://dx.doi.org/10.1109/ICSESS47205.2019.9040759]

[43] C. Chakraborttii, and H. Litz, "Explaining SSD failures using anomaly detection", *Non-Volatile Memory Workshop,* vol. 1, p. 1, 2020.

[44] M. Iaculo, F. Falanga, and O. Vitale, "Introduction to ssd", In: *Memory Mass Storage.* Springer: Berlin, Heidelberg, 2011, pp. 213-236.
[http://dx.doi.org/10.1007/978-3-642-14752-4_5]

CHAPTER 12

Profit-Based Unit Commitment Using Local and Global Search Methods

Nirmalya Mallick[1,*], **Ayani Nandi**[1] and **Ranadip Roy**[1]

[1] *Department of Electrical Engineering, Sanaka Educational Trust's Group of Institutions, Durgapur, India*

Abstract: The availability of clean energy is crucial for both the environment and human health. Numerous harmful gasses released by conventional automobiles cause illnesses and ailments in people all over the world. Nevertheless, there is growing interest in Plug-in Electric Vehicles (PEVs) to help with the energy and climate emergency. It has been noted that the manufacturing of PEVs has dramatically increased over the past ten years. The PEVs may supply the power grid with electricity while both consuming it and storing it in batteries. By effectively managing the electric demand profile and integrating electricity from PEVs into the electric grid, operating expenses can be reduced overall. The study recommends the course of action, which, in this case, is to apply the chaotic mapped Sine Cosine Algorithm advancement method and combine chaotic maps with the Harris Hawks Optimizer. It also evaluates how well the suggested better technique is implemented while taking PEVs into account.

Keywords: Chaotic mapped harris hawks optimization, Chaotic mapped sine-cosine, Plug-in electric vehicles.

INTRODUCTION

Substantial electricity demand across key sectors, encompassing residential, commercial, industrial, and agricultural domains, is driving continuous expansion of the electric power system. Currently, the electrical power industry is classified as being enormously large, highly nonlinear, and highly linked. Overloading of power grids occurs as a result of rising electricity consumption, deregulation, and privatization. Over time, the power demand has changed. In order to attain ideal scheduling, each generating unit's commitment is taken into account. Power availability towards the grid can be used to regulate power for time-varying load demands [1, 2]. Due to significant power fluctuations, the demand for electricity rises in the residential, commercial, and industrial sectors. Comparable to how real-time work in the electricity sector has been radically transformed by

* **Corresponding author Nirmalya Mallick:** Department of Electrical Engineering, Sanaka Educational Trust's Group of Institutions, Durgapur, India; E-mail: nirsayo2109@gmail.com

Santosh Kumar Das, Soumi Majumder and Nilanjan Dey (Eds.)

e-mobility [3]. It is, therefore, extremely difficult to meet the shifting demand for electricity. Every nation has a basic issue with electricity. As a developing nation, India's trajectory is significantly influenced by the extent of electricity consumption in each state, reflecting its potential for progress. Because the demand for electricity is steadily increasing, there is an urgent need to generate enough electricity to meet our energy needs. Nuclear power plants, which have high fuel prices, limited coal availability, and certain detrimental consequences on the environment and ecology, produce the majority of the electricity in India.

As a result, the coal-fired power plant must be operated in concert with various power sources, including nuclear, solar, wind, and hydropower. To achieve low production costs with high reliability, a unit commitment of plants—a phrase used to describe the coordination of plants—adopts the most conservative power plant models. The no-free-lunch theorem states that there is not a single optimization strategy present ideally suitable for every optimization problem [4]. Because of this, an associated algorithm may perform poorly on a different set of issues, even when a particular meta-heuristic method may show promising results on a range of problems. It is clear that the NFL continuously improves on existing methods and provides new meta-heuristics, which makes this study topic incredibly dynamic. Currently, unit commitment problems with various thermal unit systems are investigated and resolved using the research work recommended by Seyedali Mirjalili [5].

NOVELTY AND ROAD MAP OF THE CHAPTER

The current optimization methods show promising progress, but their research is still in preliminary stages and faces several challenges that need resolution. Uncertainties include the following:

1. Figuring out how to compute the best state that is still unattainable after obtaining local or near-optimal solutions, with a focus on Plug-in Electric Vehicles (PEVs).

2. Determining the best method for combining the benefits of characteristic enhancement designs with PEVs' ability to charge and discharge.

3. Establishing effective limitations or parameters for optimization designs that account for the behavior of PEVs.

4. Defining the stopover conditions for the full cycle of iterations in the optimization process.

5. Addressing the significant issue of lacking a well-organized and comprehensive theoretical framework involving PEVs.

Therefore, the problems are easily solved using this unique optimization technique. The proposed optimization procedure is useful for solving the most challenging issues. Moreover, the robustness of the conclusions is notably affected by the limited examination of the profit-based unit commitment problem concerning all pertinent characteristics, such as the initial status of thermal-producing units and shutdown costs. Thus, the effort is justified in continuing the suggested investigation.

LITERATURE REVIEW

The field of exploration in optimization strategies is currently a vast and active area of study where research is progressing significantly. Nowadays, researchers are engaged in various projects addressing diverse issues using multiple methods, demonstrating proficiency in accurately predicting outcomes. Ongoing research is focused on discovering new algorithms and addressing certain limitations of existing techniques. Table **1** presents survey data on heuristic approaches and meta-heuristics. Fig. (**1**) illustrates various Harris Hawks Optimizers.

Table 1. Different kinds of EVs.

Reference	Year of Publication	Research Works	Name of the Novel Technique
[6]	2021	This innovative approach was developed to predict the speed of electric vehicles while ensuring the accuracy of the vehicle's performance.	LSTM
[7]	2021	A novel methodology was devised to enhance the learning rate for optimal solutions and eliminate the non-stationary internal noise of PEVs.	ALRT-CNNs
[8]	2021	The newly implemented method aimed to determine optimal routes and provide essential information about battery recharging or electric vehicle charging stations.	GA
[9]	2021	This method was specifically crafted for multi-objective energy management in Hybrid Electric Vehicles (HEVs), focusing on energy conservation and increased electrical efficiency.	NSGA-II
[10]	2021	The innovative technique was applied to address emission dispatch and dynamic economic scheduling in the electricity market.	CSA

(Table 1) cont.....

Reference	Year of Publication	Research Works	Name of the Novel Technique
[11]	2021	A novel method was developed to control the temperature and manage the cooling of Electric Vehicles (EVs) to maintain the optimal functioning of the evaporator.	IMPC
[12]	2021	This research introduced a new method for controlling the charging of EVs within smart grid systems.	ACC
[13]	2021	A strategic approach to define the nature of charging of EVs.	FCR
[14]	2019	The IWOA method was utilized to address the challenge of identifying serviceable EV charging stations.	IWOA
[15]	2017	A collaborative charging schedule for EVs involving multiple aggregators was implemented to accommodate a higher number of EVs efficiently.	A higher overall profit was observed when compared to an equal amount of energy purchased.
[16]	2017	A Hybrid Optimization Algorithm was employed for the optimal prediction of charging stations for PEVs in distributed systems.	HOA
[17]	2017	The research paper concentrated on examining the charging system for EVs with an emphasis on solar power utilization.	Solar Powered Electric Vehicle Charging System
[18]	2017	This work focuses on the charging aggregator strategy of Evs.	Dantzig-Wolfe Decomposition [DWD]
[19]	2017	The scheduling strategy for EVs in housing demand reaction programs took into account neighbor connections.	Scheduling Strategy for Electric Vehicle
[20]	2016	Both the aggregator and customer collaborated to balance the charging scheduling of Evs, considering the impact on the smart grid.	To determine whether optimum charging schedule could solve the problem, many evolutionary optimization techniques were used.
[21]	2016	The Modified Differential Evolution method demonstrated effectiveness in offering a solution for Unit Commitment Problems while taking into account the influence of PEVs.	Modified Differential Evolution
[22]	2016	Swarm Intelligence Techniques were applied to identify the optimal power distribution outline.	

(Table 1) cont.....

Reference	Year of Publication	Research Works	Name of the Novel Technique
[23]	2015	The study demonstrated the market diffusion of EVs and included driving data.	Collecting empirical data on the market's spread of electric cars.

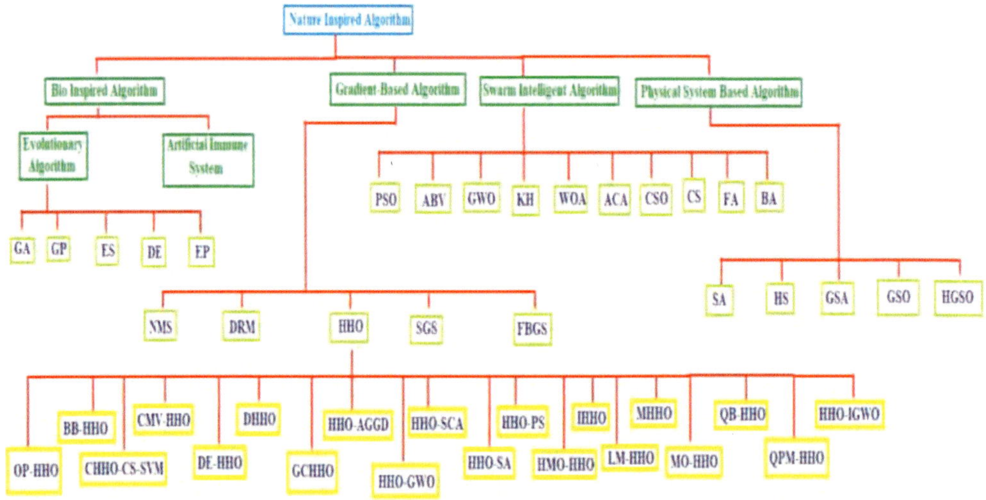

Fig. (1). Metaheuristics optimization charts.

Other existing optimization methods show promising potential for improvement, but their research remains at a preliminary stage and involves numerous challenges that need resolution. A few challenges include the best ways to leave or depart from local optima, a well-known active method for superbly consolidating the benefits of characteristic improvement designs, how to properly establish a design's bounds or restrictions, and the exciting cycle of iteration stop conditions, among others. Santosh Kumar Das *et al.* developed data mining techniques for detecting system anomalies in networks [24]. Soumi Majumder and Nilanjan Dey discuss the progression of information management, including the distribution, creation, and maintenance of knowledge [25]. Santosh Das explored various smart applications based on user specifications and requirements [26]. Santosh Das *et al.* designed a Non-linear Multi-Objective Method for hybrid peer-to-peer communications [27]. Ashish Kumar Das and Santosh Kumar Das proposed an optimal data communication system for wireless sensor networks using Artificial Neural Networks (ANN) [28]. Nilanjan Dey discussed data-driven approaches with advanced research in Machine Learning (ML) and Artificial Intelligence (AI) techniques through data analytics in healthcare and biomedical applications [29].

PROPOSED METHOD

To find an answer to the Profit-Based Unit Commitment Problem (PBUCP) in the power industry, the current study makes use of a modified version of CHHO and CSCA. This approach considers the charging and discharging nature of PEVs to determine the optimal scheduling for operating thermal units economically. The objective is to meet time-varying demand while satisfying system and physical constraint requirements. Stochastic and heuristic processes are employed to handle various operational and physical constraints associated with PBUCP.

PBUCP Spinning Reserve Constraint: Take PEVs into Account When Charging and Discharging. To meet the reserve capacity requirements for various types of power units, it is essential to take into account the minimum up and down time for each power unit. Additionally, consideration should be given to the time durations during which the ith unit remains continuously in the OFF condition. The reserve constraints must be addressed in accordance with the algorithm outlined in the flowchart provided in Fig. (**2**). The requirements for the up and down time of thermal units can be addressed through the mechanism outlined in Fig. (**3**) below.

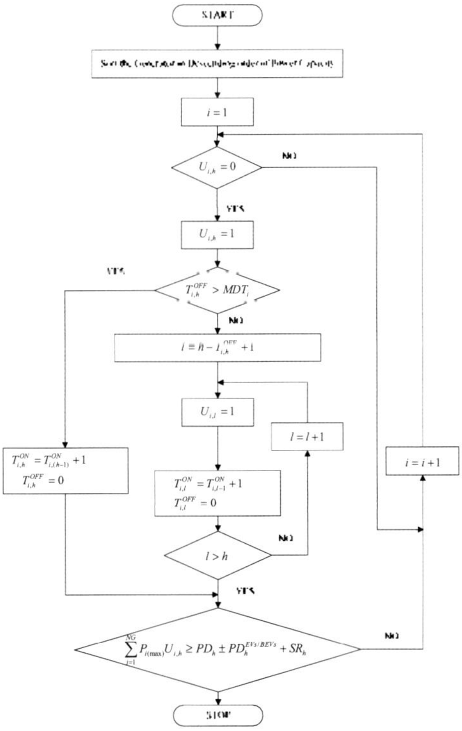

Fig. (2). Flowchart to showcase the impact of PEVs considering constraints spinning reserve.

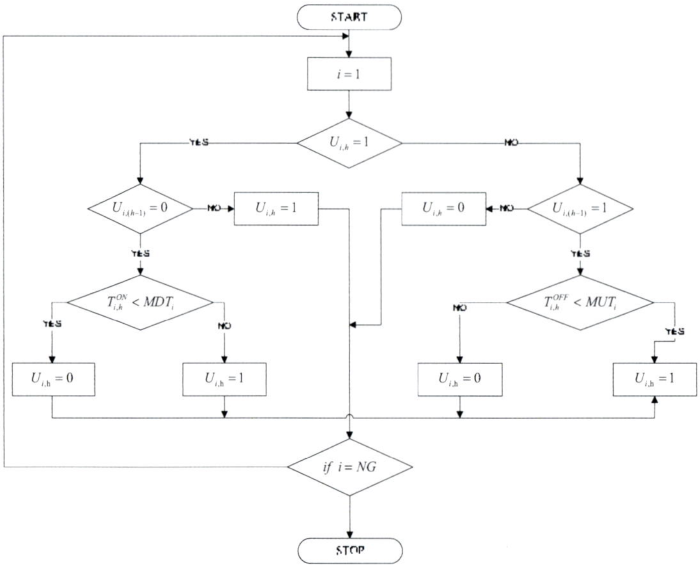

Fig. (3). Minimum up and down time flowchart.

In the analysis of complex phenomena in simulation, sampling, mathematical analysis, dynamics, and particularly heuristic optimization, the generation of random groupings over an extensive period with high consistency is crucial. From a numerical point of view, chaos is a source of randomness since it incorporates the randomness of a fundamental deterministic dynamical framework. The peculiar behavior of chaotic systems has captured the attention of various scientific research communities. Table **2** provides detailed discussions of various types of chaotic strategies. Fig. (**4**) displays the code for the modified CSCA, while Fig. (**5**) presents the modified CHHO optimization algorithm. Additionally, Fig. (**6**) illustrates the flowchart.

Initializations of random populations by $X_{K+1} = v\left[7.86X_K - 23.31X_K^2 - 28.75X_K^3 - 13.302875X_K^4\right]$

Do Estimated every search agent by objective functions

Upgraded the best values of fitness accomplished so far

Upgraded random numbers by r_1, r_2, r_3 & r_4

```
xx^0.1=rand;
u=1.07;
xx^0.1= u*(7.86*xx-23.31*(xx^2)-28.75*(xx^3)-13.302875*(xx^4));
```

if $r_4 < 0.5$ *Upgraded* locations of search agent through

$$X_m(\text{iter}+1) = X_m(\text{iter}) + r_1 \times \sin(r_2) \times |r_3 \times D_m(\text{iter}) - X_m(\text{iter})|$$

else *Upgrades* the locations of search agent through

$$X_m(\text{iter}+1) = X_m(\text{iter}) + r_1 \times \cos(r_2) \times |r_3 \times D_m(\text{iter}) - X_m(\text{iter})|$$

End While ($iter < iter_{max}$)

Returns best value of optimum solutions. Saved the values of mean fitness, standard deviations and best, worst and recorded the all-optimal assessments found over sequential trial run.

Fig. (4). Code for CSCA.

INPUTS: The population sizes are N and the nos. of maximum iterations are $iter_{MAX}$

OUTPUTS: The positions of the prey (rabbits) in adding its implication of the fitness by singer function of chaotic maps

Initializations of random populations by $X_{K-1} = v\left[7.86X_K - 23.31X_K^2 + 28.75X_K^3 - 13.302875X_K^4\right]$

While (iterations < $iter_{MAX}$)

Compute Harris hawks' significances of the fitness

Static the factor X_{Rabbit} for greatest locations of rabbits (prey)

for (all Harris hawks (X_m)), **Do** Updated energies at preliminary conditions EN_O

xx^{0,1}=rand;

u=1.07;

xx^{0,1}= u*(7.86*xx-23.31*(xx^2)+28.75*(xx^3)-13.302875*(xx^4));

if $|EN| \geq 1$ **then Positions** vectors update using $EN = 2 \times EN_0 \times \left(1 - \frac{iter}{iter_{MAX}}\right)$ →**Phases of the Exploration**

else if $b \geq 0.5$ **then** $X(iter+1) = \{X_{Hawks}^{RND}(iter) - r_1 \times abs(X_{Hawks}^{RND}(iter) - 2 \times r_2 \times X(iter))$

else if $b < 0.5$ **then**

$X(iter+1) = \{(X_{Rabbit}(iter) - X_a(iter)) - r_3 \times (Lower_B + r_4 \times (Upper_B - Lower_B))$

if $|EN| < 1$ **then** →**Phases of the Exploitation**

$X(iter+1) = \Delta X(iter) - EN \times abs(jX_{Rabbit}(iter) - X(iter))$

if (r ≥ 0.5) & $|EN| \geq 0.5$ **then** →Soft circles

Location vectors restructured by $\Delta X(iter) = (X_{Rabbit}(iter) - X(iter))$

else if (r ≥ 0.5) & $|EN| < 0.5$ **then** →**Hard circles**

Location vectors restructured by $X(iter+1) = X_{Rabbit}(iter) - EN \times abs(\Delta X(iter))$

else if (r < 0.5) & $|EN| \geq 0.5$ **then** →**Soft circles by advanced fastest dive**

Location vectors restructured by $V = X_{Rabbit}(iter) - EN \times abs(jX_{Rabbit}(iter) - X(iter))$

else if (r < 0.5) & $|EN| < 0.5$ **then** →**Hard circles by advanced fastest dive**

Location vectors restructured by $X(iter+1) = \Delta X(iter) - EN \times abs(jX_{Rabbit}(iter) - X(iter))$

end

end

end

end

Fig. (5). Code for CHHO.

Table 2. Chaotic Maps.

Sl. No.	Name of Chaotic Maps	Mathematical Formulations	Required Parameters
1	Circle	$X_{K+1} = \left\lvert X_K + b - [a - 2\pi]\sin(2\pi X_K)\right\rvert$	a = 0.5 as well as b = 0.2, it assistances to produces chaotic orders [0, 1]
2	Chebyshev	$X_{K+1} = \cos[K\cos^{-1}(X_K)]$	-

(Table 1) cont.....

Sl. No.	Name of Chaotic Maps	Mathematical Formulations	Required Parameters
3	Intermittency	$X_{K+1} = \begin{cases} \partial + X_K + CX_K & 0 < X_K \le G \\ \dfrac{X_K - G}{1 - G} & G < X_K < 1 \end{cases}$	-
4	Iterative	$X_{K+1} = \sin\left[\dfrac{a\pi}{X_K}\right]$	$a \in [0,1]$
5	Mouse or Gauss	$X_{K+1} = \begin{cases} 0 & X_K = 0 \\ 1/K \bmod[1] & otherwise \end{cases}$ $1/X_K \bmod[1] = \dfrac{1}{X_K} - \left[\dfrac{1}{X_K}\right]$	-

PERFORMANCE EVALUATION

Several standard test systems, including small, medium, and large-scale power systems with standard IEEE bus configurations, have been considered to evaluate the effectiveness of the proposed approaches in addressing the PBUCP, incorporating the impacts of PEVs and Renewable Energy Sources (RES), specifically solar power. Small-scale systems involve ten generating units, medium-scale systems comprise forty generating units, and large-scale systems consist of one hundred generating units. In this section, a comparative analysis of the algorithmic results, specifically CHHO and CSCA, is presented. To ensure proper scheduling within a designated time frame, the total number of PEVs should not exceed the maximum anticipated or registered PEVs. The Fig. in consideration, Fig. (**7**) presents data on Power in MW, encompassing Vehicle-t--Grid (V2G) and Grid-to-Vehicle (G2V) operations.

Ten Power Unit System

The efficacy of the modifier CHHO optimizer, taking the impact of PEVs during charging and discharging, is assessed using a test system. The evaluation of the CHHO technique involves 100 iterations, while both CHHO and CSCA algorithms are subjected to verification through 30 trial runs. Fig. (**8**) illustrates the convergence graph of the total profit, and the profit tables for the 10-generating unit system are presented in Table **3** for CHHO and Table **4** for CSCA.

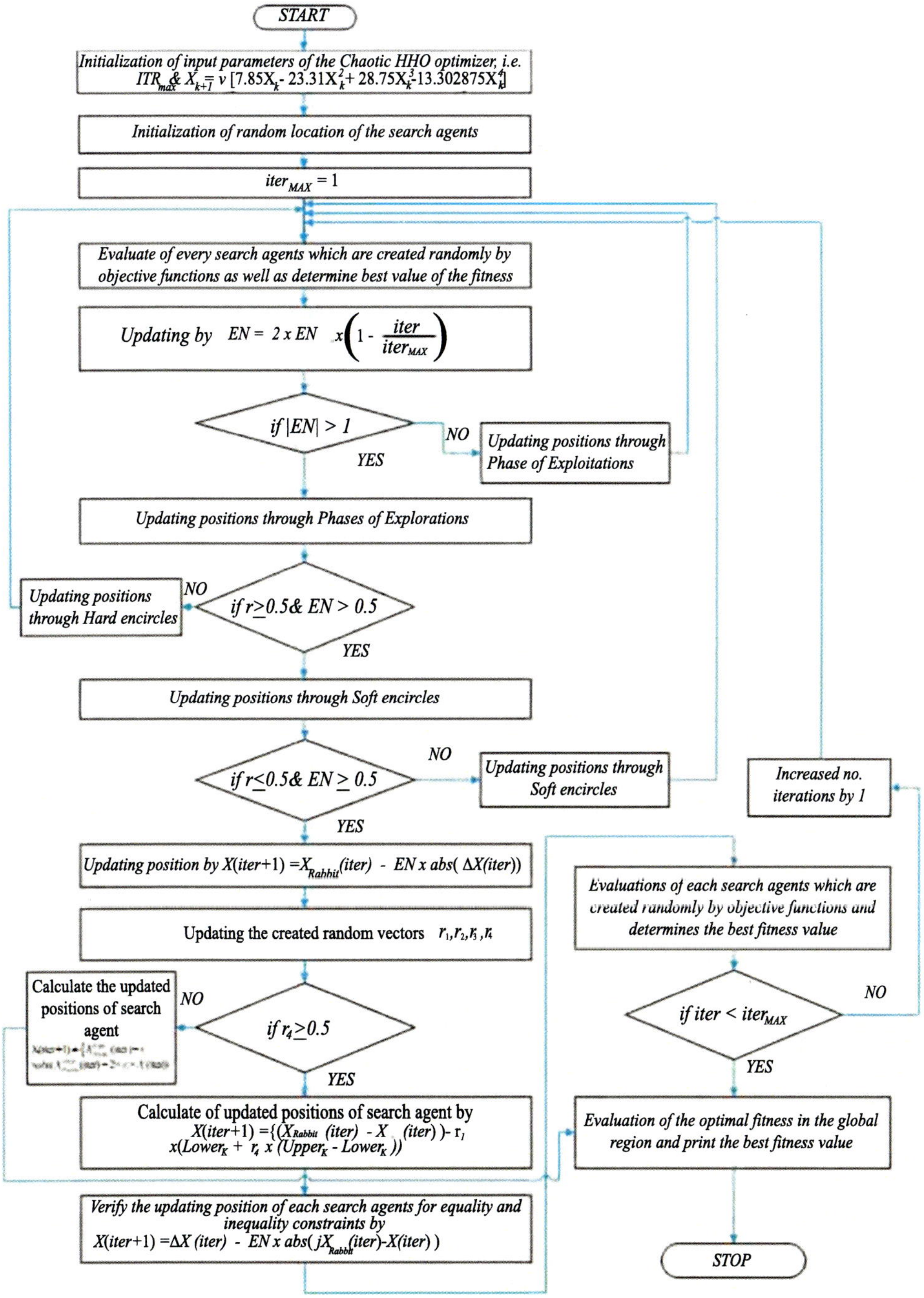

Fig. (6). Flowchart of the discussed algorithm.

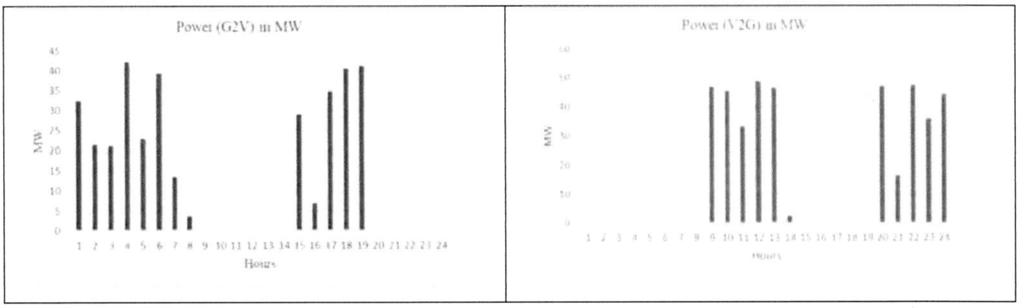

Fig. (7). Power during V2G and G2V operation.

Table 3. Profit for 10th power unit using modified CHHO.

HOURS	"Pgen1"	"Pgen2"	"Pgen3"	"Pgen4"	"Pgen5"	"Pgen6"	"Pgen7"	"Pgen8"	"Pgen9"	"Pgen10"	Hourly Profit
1	10078.25	6144.41	0	0	0	0	0	0	0	0	16222.66
2	10010	6963	0	0	0	0	0	0	0	0	16973
3	10510.5	9037.644	0	0	577.5	0	0	0	0	0	20125.644
4	10305.75	10305.75	0	0	959.2275	0	0	0	0	0	21570.728
5	10578.75	9602.0175	0	3022.5	581.25	0	0	0	0	0	23784.518
6	10442.25	10442.25	0	2983.5	900.558	0	0	0	0	0	24768.558
7	10237.5	9979.65	0	2925	562.5	0	0	0	0	0	23704.65
8	10078.25	9605.7905	0	2879.5	553.75	0	0	0	0	0	23117.291
9	10374	8737.872	2964	2964	570	0	0	0	0	0	25609.872
10	13354.25	13354.25	3815.5	3815.5	4094.912	587	733.75	0	0	0	39755.162
11	13718.25	13718.25	3919.5	3919.5	4636.1655	603	753.75	0	0	301.5	41569.916
12	14400.75	14400.75	4114.5	4114.5	4376.2455	633	791.25	316.5	0	0	43147.496
13	11193	11193	3198	3198	1482.642	492	615	246	0	0	31617.642
14	11147.5	11147.5	3185	3185	2021.985	490	612.5	0	0	0	31789.485
15	10237.5	10237.5	2925	2925	765.45	0	562.5	0	0	0	27652.95
16	10146.5	9740.863	0	2899	557.5	0	0	0	0	0	23343.863
17	10123.75	9451.8	0	2892.5	556.25	0	0	0	0	0	23024.3
18	10032.75	10032.75	0	2866.5	894.348	0	0	0	0	0	23826.348
19	10101	10101	0	2886	914.862	0	0	0	0	0	24002.862
20	10305.75	8789.106	0	2944.5	0	453	0	0	0	0	22492.356
21	10510.5	9434.502	0	3003	0	462	0	0	0	231	23641.002
22	10442.25	8894.502	0	2983.5	0	459	0	0	0	0	22779.252
23	10351.25	8851.57	0	0	0	455	0	0	0	0	19657.82
24	10260.25	6778.3045	0	0	0	0	0	0	0	0	17038.555

Table 4. Profit for 10th power unit using modified CSCA.

HOURS	"Pgen1"	"Pgen2"	"Pgen3"	"Pgen4"	"Pgen5"	"Pgen6"	"Pgen7"	"Pgen8"	"Pgen9"	"Pgen10"	Hourly Profit
1	10078.25	6144.41	0	0	0	0	0	0	0	0	16222.66
2	10010	6963	0	0	0	0	0	0	0	0	16973
3	10510.5	9384.144	0	0	0	0	0	231	0	0	20125.644
4	10305.75	8093.9775	0	2944.5	0	0	0	0	226.5	0	21570.728
5	10578.75	9602.0175	0	3022.5	581.25	0	0	0	0	0	23784.518
6	10442.25	10442.25	0	2983.5	900.558	0	0	0	0	0	24768.558
7	10237.5	9979.65	0	2925	562.5	0	0	0	0	0	23704.65
8	10078.25	9605.7905	0	2879.5	553.75	0	0	0	0	0	23117.291
9	10374	10374	0	2964	1441.872	456	0	0	0	0	25609.872
10	13354.25	13354.25	3815.5	3815.5	4094.912	587	733.75	0	0	0	39755.162
11	13718.25	13718.25	3919.5	3919.5	4636.1655	603	753.75	301.5	0	0	41569.916
12	14400.75	14400.75	4114.5	4114.5	4376.2455	633	791.25	316.5	0	0	43147.496
13	11193	11193	3198	3198	1728.642	492	615	0	0	0	31617.642
14	11147.5	11147.5	3185	3185	2021.985	490	612.5	0	0	0	31789.485
15	10237.5	10237.5	2925	2925	877.95	450	0	0	0	0	27652.95
16	10146.5	9740.863	0	2899	557.5	0	0	0	0	0	23343.863
17	10123.75	9451.8	0	2892.5	556.25	0	0	0	0	0	23024.3
18	10032.75	10032.75	0	2866.5	894.348	0	0	0	0	0	23826.348
19	10101	10101	0	2886	914.862	0	0	0	0	0	24002.862
20	10305.75	8675.856	0	2944.5	566.25	0	0	0	0	0	22492.356
21	10510.5	9550.002	0	3003	577.5	0	0	0	0	0	23641.002
22	10442.25	8779.752	0	2983.5	573.75	0	0	0	0	0	22779.252
23	10351.25	6349.07	0	2957.5	0	0	0	0	0	0	19657.82
24	10260.25	6778.3045	0	0	0	0	0	0	0	0	17038.555

Forty Power Unit System

Hourly profit tables for 40 units are outlined in Tables **5a** and **5b** for the modified CHHO optimization algorithm and Tables **6a** and **6b** for the modified CSCA optimization algorithm.

Table 5a. For 40th power unit hourly profit (profile for modified CHHO) (Unit 1 – Unit-20).

HOURS	Pgen 1	Pgen 2	Pgen 3	Pgen 4	Pgen 5	Pgen 6	Pgen 7	Pgen 8	Pgen 9	Pgen 10	Pgen 11	Pgen 12	Pgen 13	Pgen 14	Pgen 15	Pgen 16	Pgen 17	Pgen 18	Pgen 19	Pgen 20
1	10078.25	5247.335	0	0	0	0	0	0	0	0	10078.25	5247.335	0	0	0	0	0	0	0	0
2	10010	5656.75	0	0	0	0	0	0	0	0	10010	5656.75	0	0	0	0	0	0	0	0
3	10510.5	8251.089	0	0	0	0	0	0	0	0	10510.5	8251.089	0	0	0	0	0	0	0	0
4	10305.75	9046.69313	0	0	566.25	0	566.25	0	0	0	10305.75	9046.6931	0	0	0	0	0	0	0	0
5	10578.75	9747.62063	0	3022.5	581.25	465	581.25	0	0	0	10578.75	9747.6206	0	3022.5	0	465	0	0	0	0
6	10442.25	9700.7355	0	2983.5	573.75	459	573.75	0	0	0	10442.25	9700.7355	0	2983.5	0	459	0	0	0	0
7	10237.5	8333.2125	0	2925	562.5	450	0	0	0	0	10237.5	8333.2125	0	2925	0	450	0	0	0	0
8	10078.25	7538.36488	2879.5	2879.5	553.75	443	0	0	0	0	10078.25	7538.3649	0	2879.5	0	443	0	0	0	0

(Table 5a) cont.....

HOURS	Pgen 1	Pgen 2	Pgen 3	Pgen 4	Pgen 5	Pgen 6	Pgen 7	Pgen 8	Pgen 9	Pgen 10	Pgen 11	Pgen 12	Pgen 13	Pgen 14	Pgen 15	Pgen 16	Pgen 17	Pgen 18	Pgen 19	Pgen 20
9	10374	9814.03201	2963.99999	2964	570	456	0	0	0	0	10374	9814.032	2964	2964	570	456	0	0	228	0
10	13354.25	13354.25	3815.5	3815.5	4754.7	1228.8845	733.75	293.5	293.5	0	13354.25	13354.25	3815.5	3815.5	4754.7	1228.8845	733.75	293.5	0	0
11	13718.25	13718.25	3919.5	3919.5	4884.3	1531.84612	753.75	301.5	301.5	0	13718.25	13718.25	3919.5	3919.5	4884.3	1531.8461	753.75	301.5	0	0
12	14400.75	14400.75	4114.5	4114.5	5127.3	1651.57612	791.25	316.5	316.5	0	14400.75	14400.75	4114.5	4114.5	5127.3	1651.5761	791.25	316.5	316.5	0
13	11193	11193	3198	3198	3165.589	492	615	0	0	0	11193	11193	3198	3198	3165.5895	492	615	0	0	0
14	11147.5	11147.5	3185	3185	2557.003	490	612.5	0	0	0	11147.5	11147.5	3185	3185	2557.0038	490	0	0	0	0
15	10237.5	10237.5	2925	2925	1749.262	0	0	0	0	0	10237.5	10237.5	2925	2925	1749.2625	0	0	225	0	0
16	10146.5	9105.03425	2899	2899	557.5	0	0	0	0	0	10146.5	9105.0343	0	2899	557.5	0	0	0	0	0
17	10123.75	8177.9875	2892.5	2892.5	556.25	0	0	0	0	0	10123.75	8177.9875	0	2892.5	0	0	0	0	0	0
18	10487.75	9620.148	0	2996.5	576.25	461	0	230.5	0	0	10487.75	9620.148	0	2996.5	0	0	0	230.5	0	0
19	10101	9428.2845	0	2886	555	444	0	0	0	0	10101	9428.2845	0	2886	0	0	0	0	0	0
20	10305.75	9835.536	0	0	0	453	0	226.5	0	226.5	10305.75	9835.536	0	2944.5	0	0	0	0	0	0
21	10510.5	10490.7495	0	0	0	462.000001	577.5	0	0	0	10510.5	10490.75	0	0	0	462	0	231	0	0
22	10442.25	10442.25	0	0	0	616.437	573.75	0	0	0	10442.25	10442.25	0	0	0	616.437	0	0	0	0
23	10351.25	10092.81	2957.5	0	0	0	568.75	0	0	0	10351.	10092.81	0	0	568.75	455	0	0	0	0
24	10260.25	10260.25	2931.5	0	0	0	0	0	0	0	10260.25	10260.25	0	0	1008.0978	0	563.75	225.5	0	0

Table 5b. 40th power unit hourly profit (profile for modified CHHO) (Unit 21- Unit 40).

HOURS	Pgen 21	Pgen 22	Pgen 23	Pgen 24	Pgen 25	Pgen 26	Pgen 27	Pgen 28	Pgen 29	Pgen 30	Pgen 31	Pgen 32	Pgen 33	Pgen 34	Pgen 35	Pgen 36	Pgen 37	Pgen 38	Pgen 39	Pgen 40
1	10078.25	5247.335	0	0	0	0	0	0	0	0	10078.25	5247.335	0	0	0	0	0	0	0	0
2	10010	5656.75	0	2860	0	0	0	0	0	0	10010	5656.75	0	0	0	0	0	0	0	0
3	10510.5	8251.089	0	3003	0	0	0	0	0	0	10510.5	8251.089	0	0	0	0	0	0	0	0
4	10305.75	9046.69313	0	2944.5	0	0	0	0	0	0	10305.75	9046.6931	0	0	0	0	0	0	0	0
5	10578.75	9747.62063	0	3022.5	0	0	0	0	0	0	10578.75	9747.6206	0	0	0	0	0	0	0	0
6	10442.25	9700.7355	0	2983.5	0	0	0	0	0	0	10442.25	9700.7355	0	2983.5	0	0	0	0	0	0
7	10237.5	8333.2125	2925	2925	0	0	0	0	0	0	10237.5	8333.2125	2925	2925	0	0	0	0	0	0
8	10078.25	7538.36488	2879.5	2879.5	0	0	0	0	0	0	10078.25	7538.3649	2879.5	2879.5	0	0	0	0	0	0
9	10374	9814.03201	2963.99999	2964	0	0	570	0	0	0	10374	9814.032	2964	2964	0	456	0	0	0	0
10	13354.25	13354.25	3815.5	3815.5	4754.7	1228.8845	733.75	293.5	0	0	13354.25	13354.25	3815.5	3815.5	4754.7	1228.8845	733.75	293.5	0	0
11	13718.25	13718.25	3919.5	3919.5	4884.3	1531.84612	753.75	301.5	0	0	13718.25	13718.25	3919.5	3919.5	4884.3	1531.8461	753.75	301.5	0	0
12	14400.75	14400.75	4114.5	4114.5	5127.3	1651.57612	791.25	316.5	0	0	14400.75	14400.75	4114.5	4114.5	5127.3	1651.5761	791.25	316.5	0	0
13	11193	11193	3198	3198	3165.5895	492	615	0	0	0	11193	11193	3198	3198	3165.5895	492	615	0	0	0
14	11147.5	11147.5	3185	3185	2557.00375	490	0	0	0	0	11147.5	11147.5	3185	3185	2557.0038	490	0	0	0	0
15	10237.5	10237.5	0	2925	1749.2625	450.000006	0	225	0	0	10237.5	10237.5	0	2925	1749.2625	0	0	0	0	0
16	10146.5	9105.03425	0	2899	0	0	0	0	0	0	10146.5	9105.0343	0	2899	0	0	0	0	0	0
17	10123.75	8177.9875	0	2892.5	0	0	0	0	0	0	10123.75	8177.9875	0	2892.5	0	0	0	0	0	0
18	10487.75	9620.148	0	2996.5	0	0	0	0	230.5	0	10487.75	9620.148	0	2996.5	0	0	576.25	0	0	230.5
19	10101	9428.2845	0	2886	0	0	0	222	0	0	10101	9428.2845	0	2886	0	0	555	0	0	0
20	10305.75	9835.536	2944.5	2944.5	0	453	566.25	0	0	0	10305.75	9835.536	0	2944.5	0	453	566.25	0	0	0
21	10510.5	10490.7495	3003	3003	0	462.000001	577.5	0	0	231	10510.5	10490.75	0	3003	0	462	0	0	0	0
22	10442.25	10442.25	2983.5	2983.5	0	616.437	573.75	0	0	229.5	10442.25	10442.25	0	2983.5	0	616.437	0	0	229.5	0
23	10351.25	10092.81	2957.5	0	0	0	0	0	0	0	10351.25	0	2957.5	0	568.75	0	0	0	0	0
24	10260.25	0	2931.5	0	0	0	0	0	0	0	10260.25	0	2931.5	0	1008.0978	0	0	0	0	0

Table 6a. For 40th power unit hourly profit (profile for modified CSSA) (Unit 1 – Unit-20).

HOURS	Pgen 1	Pgen 2	Pgen 3	Pgen 4	Pgen 5	Pgen 6	Pgen 7	Pgen 8	Pgen 9	Pgen 10	Pgen 11	Pgen 12	Pgen 13	Pgen 14	Pgen 15	Pgen 16	Pgen 17	Pgen 18	Pgen 19	Pgen 20
1	10078.25	5247.335	0	0	0	0	0	0	0	0	10078.25	5247.335	0	0	0	0	0	0	0	0
2	10010	5656.75	2860	0	0	0	0	0	0	0	10010	5656.75	0	0	0	0	0	0	0	0
3	10510.5	8251.089	3003	0	0	0	0	0	0	0	10510.5	8251.089	0	0	0	0	0	0	0	0
4	10305.75	9046.6931	2944.5	0	0	453	0	0	0	0	10305.75	9046.6931	0	0	0	0	0	0	226.5	0
5	10578.75	9166.3706	3022.5	3022.5	0	465	0	0	0	0	10578.75	9166.3706	0	0	0	0	0	0	0	0
6	10442.25	9729.423	2983.5	2983.5	573.75	459	0	0	0	0	10442.25	9729.423	0	0	0	0	0	0	0	0
7	10237.5	9626.9625	2925	2925	562.5	0	0	0	0	0	10237.5	9626.9625	0	0	562.5	0	0	0	0	0
8	10078.25	8811.9899	2879.5	2879.5	553.75	0	0	0	0	0	10078.25	8811.9899	0	2879.5	553.75	0	0	0	0	0
9	10374	10374	2964	2964	811.376	0	0	0	0	0	10374	10374	2964	2964	811.376	0	570	0	0	0
10	13354.25	13354.25	3815.5	3815.5	4754.7	1228.8845	733.75	293.5	293.5	0	13354.25	13354.25	3815.5	3815.5	4754.7	1228.8845	733.75	293.5	0	0

(Table 6a) cont.....

HOURS	Pgen 1	Pgen 2	Pgen 3	Pgen 4	Pgen 5	Pgen 6	Pgen 7	Pgen 8	Pgen 9	Pgen 10	Pgen 11	Pgen 12	Pgen 13	Pgen 14	Pgen 15	Pgen 16	Pgen 17	Pgen 18	Pgen 19	Pgen 20
11	13718.25	13718.25	3919.5	3919.5	4884.3	1531.8461	753.75	301.5	301.5	0	13718.25	13718.25	3919.5	3919.5	4884.3	1531.8461	753.75	301.5	0	0
12	14400.75	14400.75	4114.5	4114.5	5127.3	1651.5761	791.25	316.5	316.5	0	14400.75	14400.75	4114.5	4114.5	5127.3	1651.5761	791.25	316.5	0	316.5
13	11193	11193	3198	3198	3165.5895	492	615	0	0	0	11193	11193	3198	3198	3165.5895	492	615	0	0	0
14	11147.5	11147.5	3185	3185	2557.0038	490	0	0	0	0	11147.5	11147.5	3185	3185	2557.0038	490	612.5	0	0	0
15	10237.5	10186.763	2925	2925	0	0	0	0	0	0	10237.5	10186.763	2925	2925	562.5	0	562.5	0	0	0
16	10146.5	7794.9093	2899	2899	0	0	0	0	0	0	10146.5	7794.9093	2899	2899	0	0	0	0	0	0
17	10123.75	6870.8	2892.5	2892.5	0	0	0	0	0	0	10123.75	6870.8	2892.5	2892.5	0	0	0	0	0	0
18	10487.75	8582.898	2996.5	2996.5	0	0	0	0	0	0	10487.75	8582.898	2996.5	2996.5	0	0	0	0	0	0
19	10101	8262.7845	2886	2886	0	0	0	222	0	0	10101	8262.7845	2886	2886	0	0	0	0	0	0
20	10305.75	8759.661	2944.5	2944.5	0	453	0	0	0	0	10305.75	8759.661	2944.5	2944.5	0	0	0	0	0	0
21	10510.5	8758.2495	3003	3003	0	462	0	0	0	0	10510.5	8758.2495	3003	3003	0	0	0	0	0	0
22	10442.25	9394.812	0	2983.5	0	459	573.75	229.5	0	0	10442.25	9394.812	2983.5	2983.5	0	0	0	0	0	0
23	10351.25	9759.295	0	0	0	0	568.75	0	0	0	10351.25	9759.295	0	0	0	0	568.75	0	0	0
24	10260.25	9955.1485	0	0	0	0	563.75	0	0	0	10260.25	9955.1485	0	0	0	0	563.75	0	0	0

Table 6b. Profile for hourly profit using modified CSCA (Unit 21- Unit 40).

HOURS	Pgen 21	Pgen 22	Pgen 23	Pgen 24	Pgen 25	Pgen 26	Pgen 27	Pgen 28	Pgen 29	Pgen 30	Pgen 31	Pgen 32	Pgen 33	Pgen 34	Pgen 35	Pgen 36	Pgen 37	Pgen 38	Pgen 39	Pgen 40
1	10078.25	5247.335	0	0	0	0	0	0	0	0	10078.25	5247.335	0	0	0	0	0	0	0	0
2	10010	5656.75	0	0	0	0	0	0	0	0	10010	5656.75	0	0	0	0	0	0	0	0
3	10510.5	8251.089	0	0	0	0	0	0	0	0	10510.5	8251.089	0	0	0	0	0	0	0	0
4	10305.75	9046.6931	0	0	0	0	0	0	0	0	10305.75	9046.6931	0	0	0	453	0	0	0	0
5	10578.75	9166.3706	3022.5	3022.5	0	465	0	0	0	0	10578.75	9166.3706	0	0	0	465	0	0	0	0
6	10442.25	9729.423	2983.5	2983.5	0	459	0	0	0	0	10442.25	9729.423	0	0	0	459	0	0	0	0
7	10237.5	9626.9625	2925	2925	0	450	0	0	0	0	10237.5	9626.9625	0	0	562.5	0	0	0	0	0
8	10078.25	8811.9899	2879.5	2879.5	0	443	0	0	0	0	10078.25	8811.9899	0	0	553.75	0	0	0	0	0
9	10374	10374	2964	2964	0	456	0	0	0	0	10374	10374	0	2964	811.376	0	570	0	0	0
10	13354.25	13354.25	3815.5	3815.5	4754.7	1228.8845	733.75	293.5	0	0	13354.25	13354.25	3815.5	3815.5	4754.7	1228.8845	733.75	293.5	0	0
11	13718.25	13718.25	3919.5	3919.5	4884.3	1531.8461	753.75	301.5	0	0	13718.25	13718.25	3919.5	3919.5	4884.3	1531.8461	753.75	301.5	0	0
12	14400.75	14400.75	4114.5	4114.5	5127.3	1651.5761	791.25	316.5	0	0	14400.75	14400.75	4114.5	4114.5	5127.3	1651.5761	791.25	316.5	0	0
13	11193	11193	3198	3198	3165.5895	492	615	0	0	0	11193	11193	3198	3198	3165.5895	492	615	0	0	0
14	11147.5	11147.5	3185	3185	2557.0038	490	0	0	0	0	11147.5	11147.5	3185	3185	2557.0038	490	0	0	0	0
15	10237.5	10186.763	2925	2925	562.5	0	0	0	0	0	10237.5	10186.763	2925	2925	562.5	0	0	0	0	0
16	10146.5	7794.9093	2899	2899	557.5	0	0	0	0	0	10146.5	7794.9093	0	2899	0	0	0	0	0	0
17	10123.75	6870.8	2892.5	2892.5	0	0	0	0	0	0	10123.75	6870.8	0	2892.5	0	0	0	0	0	0
18	10487.75	8582.898	0	2996.5	0	461	0	0	230.5	0	10487.75	8582.898	0	2996.5	0	0	0	0	0	0
19	10101	8262.7845	0	2886	0	444	0	0	0	0	10101	8262.7845	0	2886	0	0	0	0	0	0
20	10305.75	8759.661	0	2944.5	0	453	0	0	0	0	10305.75	8759.661	0	2944.5	0	453	0	0	0	0
21	10510.5	8758.2495	0	3003	0	462	0	0	0	0	10510.5	8758.2495	0	3003	0	462	0	0	0	0
22	10442.25	9394.812	0	2983.5	0	0	573.75	0	0	0	10442.25	9394.812	0	2983.5	0	459	0	0	0	0
23	10351.25	9759.295	0	0	568.75	0	568.75	0	0	0	10351.25	9759.295	0	0	0	0	0	0	0	0
24	10260.25	9955.1485	0	0	563.75	0	563.75	0	0	0	10260.25	0	0	0	0	0	0	0	0	0

Hundred Power Unit System

Figs (**8a** and **8b**) display the hourly profit table for 100 units using modified CHHO, while Figs (**8c** and **8d**) present the same for modified CSCA. Hypothesis tests are also conducted to evaluate the performance. The convergence graph of total profit for the 10th, 40th, and 100th power units using both CHHO and CSCA is illustrated in Fig. (**9**).

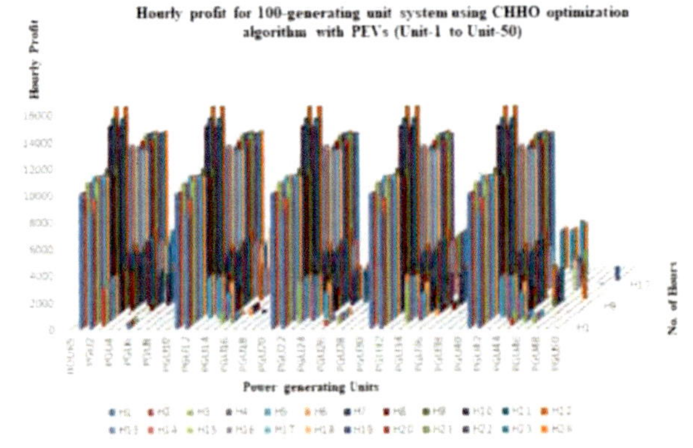

Fig. (8a). Hourly profit of 100th unit (Unit-1 to Unit-50).

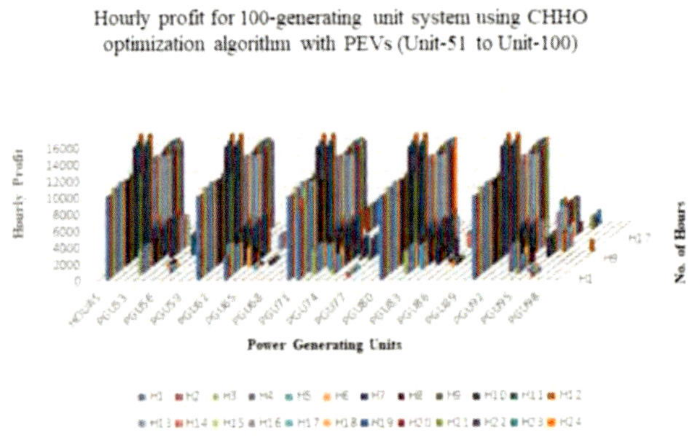

Fig. (8b). Profile for 100th power unit (Unit-51 to Unit-100).

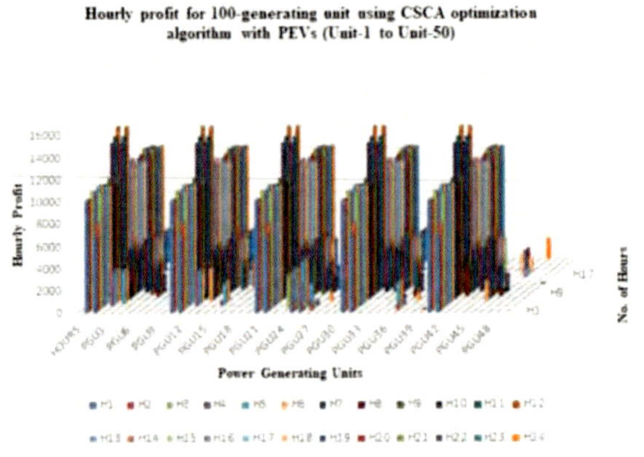

Fig. (8c). Hourly profit of 100th power unit (Unit-1 to Unit-50).

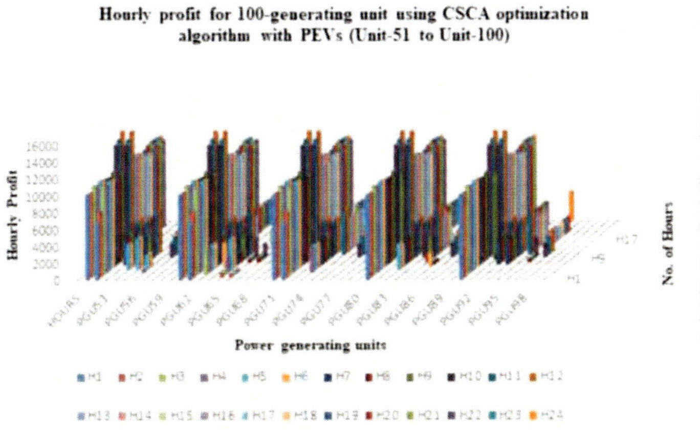

Fig. (8d). Profile for 100th power unit (Unit-51 to Unit-100).

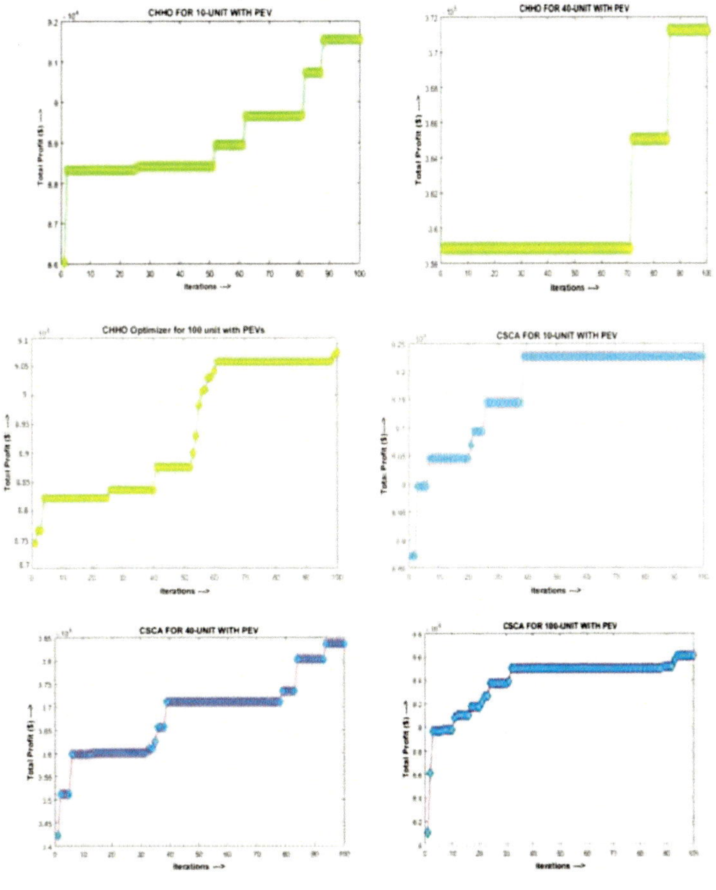

Fig. (9). Convergence for 10th, 40th and 100th power unit.

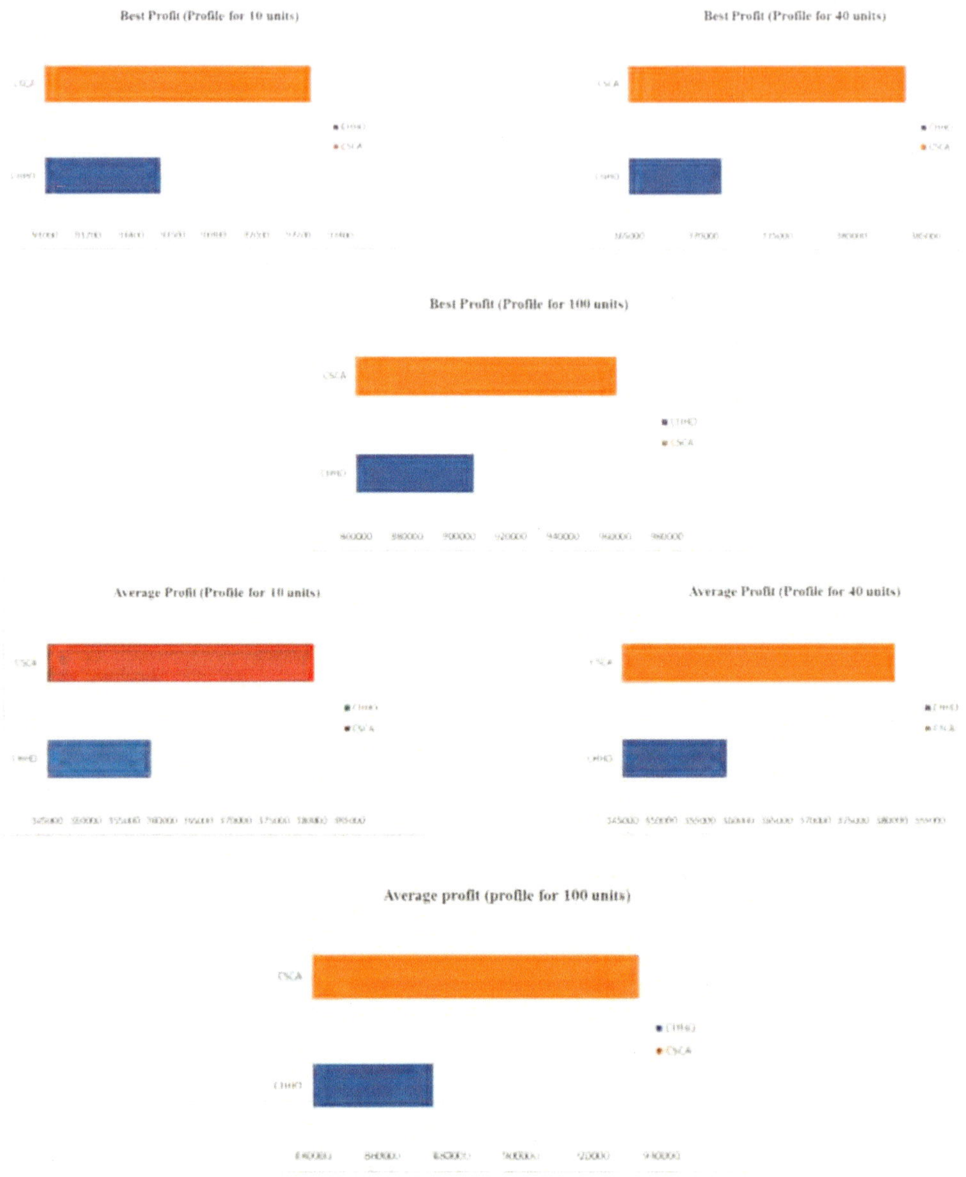
Best Profit (Profile for 10 units)
Best Profit (Profile for 40 units)
Best Profit (Profile for 100 units)
Average Profit (Profile for 10 units)
Average Profit (Profile for 40 units)
Average profit (profile for 100 units)

(Fig. 10) contd.....

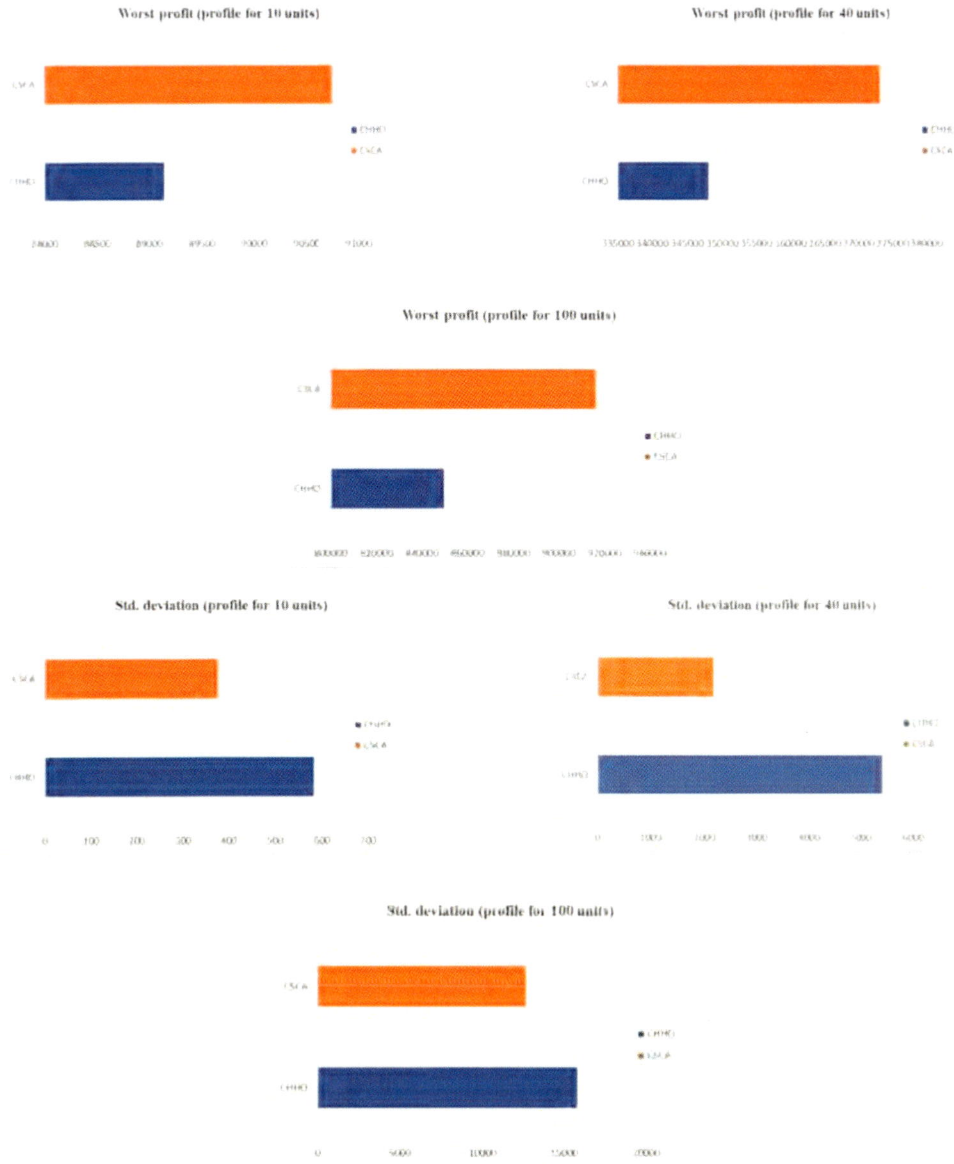
Worst profit (profile for 10 units)
Worst profit (profile for 40 units)
Worst profit (profile for 100 units)
Std. deviation (profile for 10 units)
Std. deviation (profile for 40 units)
Std. deviation (profile for 100 units)

(Fig. 10) contd.....

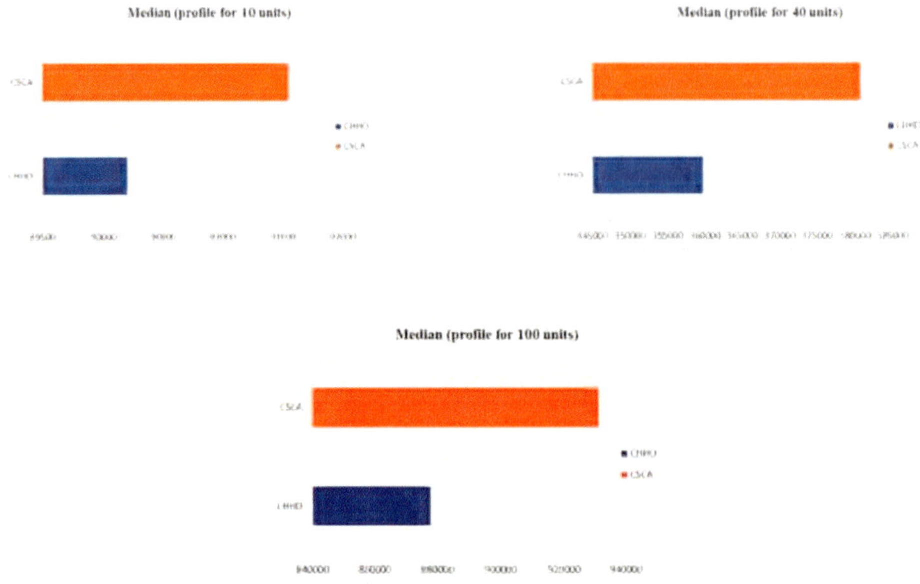

Fig. (10). Statistical results

Fig. (**10**) shows the comparative statistics for profit, along with standard deviation and median values of solutions, over 30 trial outcomes for PBUCP. The hypothetical test results are displayed in Fig. (**11**).

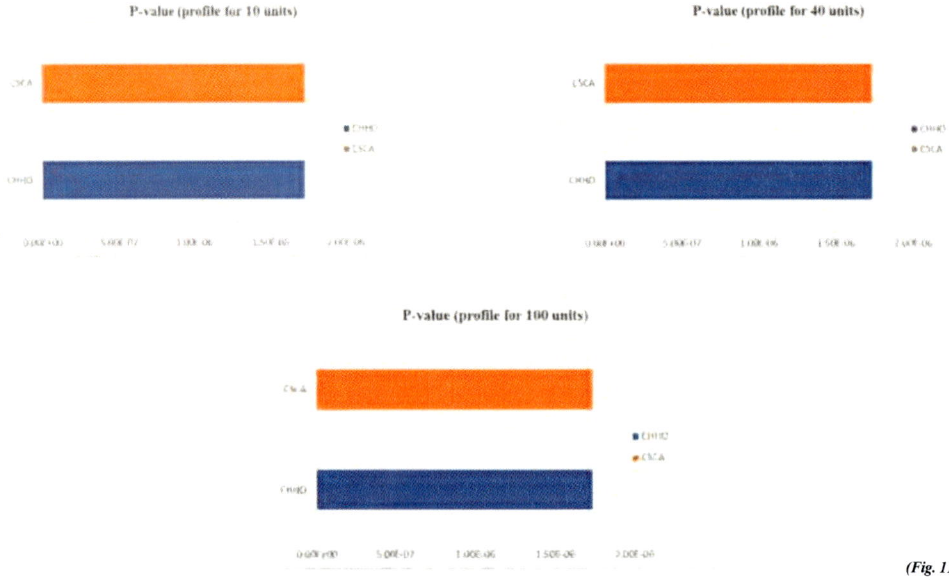

(Fig. 11) contd.....

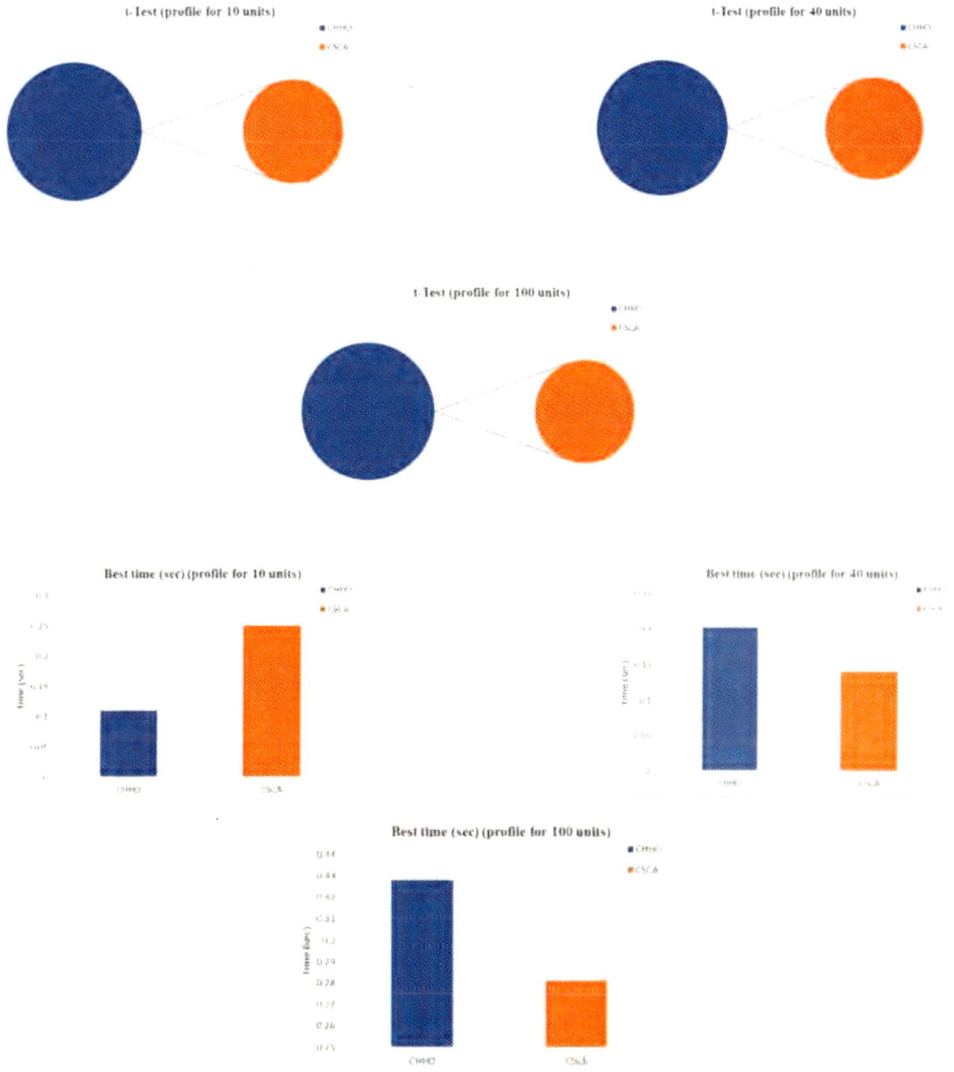

Fig. (11). Hypothetical testing results.

CONCLUSION

The PBUCP has been successfully addressed using modified CHHO and modified CSCA, taking into account the impact of PEVs. Thermal power units for small, medium, and large power sectors were used in a conventional test system to evaluate the performance of the suggested hybrid optimizer. The effectiveness of the algorithm was verified across different scenarios, including 10 power units, 40 power units, and 100 power units. Scheduling for the 10th, 40th, and 100th power units demonstrated effective utilization and improved profit values using the suggested optimizers. When the CSCA technique was used to simulate results for profit, the simulated results outperformed other known and emerging meta-heuristics, heuristics, and evolutionary search optimizers. This strong optimizer has the potential to successfully handle profit-based unit commitment problems in contemporary power industries. A thorough examination of the range of profit values was carried out, including the average, best, and worst values as well as the standard deviation and median values. The t-test and the Wilcoxon rank sum method were used in hypothesis testing to calculate p- and h-values. The best, medium, and worst timings for simulations were taken into consideration while examining computational times.

REFERENCES

[1] R.H. Kerr, "Unit commitment, IEEE transactions on power apparatus and system",

[2] C.J. Baldwin, K.M. Dale, R.F. Dittrich, and A. Study, "Of economic shutdown of generating units in daily dispatch, aiee transaction of power apparatus and systems", *Vol. PAS,* vol. 78, pp. 1272-1284, 1959.

[3] K.D. Lee, "“R,” H", *Vierra. G.D. Nagel R.T. Jenkins,* vol. 104, no. 8, pp. 2072-2078, 1985.

[4] D.H. Wolpert, and W.G. Macready, "No free lunch theorems for optimization", *IEEE Trans. Evol. Comput.,* vol. 1, no. 1, pp. 67-82, 1997. [http://dx.doi.org/10.1109/4235.585893]

[5] S. Mirjalili, "SCA: A sine cosine algorithm for solving optimization problems", *Knowl. Base. Syst.,* vol. 96, pp. 120-133, 2016. [http://dx.doi.org/10.1016/j.knosys.2015.12.022]

[6] Y.N. Malek, M. Najib, M. Bakhouya, and M. Essaaidi, "Multivariate deep learning approach for electric vehicle speed forecasting", *Big Data Mining and Analytics,* vol. 4, no. 1, pp. 56-64, 2021. [http://dx.doi.org/10.26599/BDMA.2020.9020027]

[7] H. Huang, J. Wu, T.C. Lim, M. Yang, and W. Ding, "Pure electric vehicle nonstationary interior sound quality prediction based on deep CNNs with an adaptable learning rate tree", *Mech. Syst. Signal Process.,* vol. 148, p. 107170, 2021. [http://dx.doi.org/10.1016/j.ymssp.2020.107170]

[8] S. Karakatič, "Optimizing nonlinear charging times of electric vehicle routing with genetic algorithm", In: *Expert Syst. Appl.* vol. 164. , 2021, p. 114039. [http://dx.doi.org/10.1016/j.eswa.2020.114039]

[9] Y. Li, S. Wang, X. Duan, S. Liu, J. Liu, and S. Hu, "Multi-objective energy management for Atkinson cycle engine and series hybrid electric vehicle based on evolutionary NSGA-II algorithm using digital twins", *Energy Convers. Manage.,* vol. 230, p. 113788, 2021. [http://dx.doi.org/10.1016/j.enconman.2020.113788]

[10] Y. Zou, J. Zhao, D. Ding, F. Miao, and B. Sobhani, "Solving dynamic economic and emission dispatch in power system integrated electric vehicle and wind turbine using multi-objective virus colony search algorithm", *Sustain Cities Soc.,* vol. 67, p. 102722, 2021. [http://dx.doi.org/10.1016/j.scs.2021.102722]

[11] Y. Xie, "An improved intelligent model predictive controller for cooling system of electric vehicle", *Appl. Therm. Eng.,* vol. 182, p. 116084, 2021. [http://dx.doi.org/10.1016/j.applthermaleng.2020.116084]

[12] A. Al Zishan, M.M. Haji, and O.A. Member, *Adaptive congestion control for electric vehicle charging in the smart grid,* vol. 3053, no. c, 2021.

[http://dx.doi.org/10.1109/TSG.2021.3051032]

[13] P. Astero, and C. Evens, "Optimum day-ahead bidding profiles of electrical vehicle charging stations in FCR markets", *Electr. Power Syst. Res.*, vol. 190, p. 106667, 2021. [http://dx.doi.org/10.1016/j.epsr.2020.106667]

[14] H. Zhang, L. Tang, C. Yang, and S. Lan, "Locating electric vehicle charging stations with service capacity using the improved whale optimization algorithm", *Adv. Eng. Inform.*, vol. 41, no. January, p. 100901, 2019. [http://dx.doi.org/10.1016/j.aei.2019.02.006]

[15] V. Gupta, "Multi-aggregator collaborative electric vehicle charge scheduling (cevcs) under variable energy purchase and ev cancellation events", In: *IEEE Transactions on Industrial Informatics* vol. 3203. , 2017, pp. 1-9. [http://dx.doi.org/10.1109/TII.2017.2778762]

[16] A. Awasthi, K. Venkitusamy, S. Padmanaban, R. Selvamuthukumaran, F. Blaabjerg, and A.K. Singh, "Optimal planning of electric vehicle charging station at the distribution system using hybrid optimization algorithm", *Energy,* vol. 133, pp. 70-78, 2017. [http://dx.doi.org/10.1016/j.energy.2017.05.094]

[17] S. Khan, A. Ahmad, F. Ahmad, and M.S. Shemami, "A comprehensive review on solar powered electric vehicle charging system", *Smart Sci.,* vol. 0477, no. December, pp. 1-26, 2017. [http://dx.doi.org/10.1080/23080477.2017.1419054]

[18] M.H. Amini, G.S. Member, P. Mcnamara, and P. Weng, *Hierarchical electric vehicle charging aggregator strategy using dantzig-wolfe decomposition.,* vol. 2356, no. c, pp. 1-9, 2017. [http://dx.doi.org/10.1109/MDAT.2017.2759505]

[19] S. Pal, S. Member, R. Kumar, and S. Member, *Electric vehicle scheduling strategy in residential demand response programs with neighbor,* vol. 3203, no. c, 2017. [http://dx.doi.org/10.1109/TII.2017.2787121]

[20] P. Goyal, A. Sharma, S. Vyas, and R. Kumar, *Customer and aggregator balanced dynamic electric vehicle charge scheduling in a smart grid framework.*, 2016, pp. 276-283. [http://dx.doi.org/10.1109/ICEPES.2016.7915943]

[21] N. K. Navin, A modified differential evolution approach to PHEV integrated thermal unit commitment proble," 2016 IEEE 7th Power India International Conference (PIICON), Bikaner, India, pp. 1-4,2016. [http://dx.doi.org/10.1109/POWERI.2016.8077284]

[22] I. Rahman, P. M. Vasant, B. Singh, and M. Singh, "Optimal power allocation scheme for plug-in hybrid electric vehicles using swarm intelligence techniques", *Cogent Engineering,* vol. 3, no. 1, 2016. [http://dx.doi.org/10.1080/23311916.2016.1203083]

[23] T. Gnann, P. Plötz, A. Kühn, and M. Wietschel, "Modelling market diffusion of electric vehicles with real world driving data German market and policy options", *Transp. Res.* , vol. 77, pp. 95-112, 2020. [http://dx.doi.org/10.1016/j.tra.2015.04.001]

[24] "Santosh Kumar Das and Nilanjan Dey", In: *Constraint Decision-Making Systems in Engineering.* IGI Global, 2023. [http://dx.doi.org/10.4018/978-1-6684-7343-6]

[25] "Soumi Majumder and Nilanjan Dey", *AI-empowered Knowledge Management,* vol. 107, 2022. [http://dx.doi.org/10.1007/978-981-19-0316-8]

[26] Santosh Das, "Smart Design and Its Applications: Challenges and Techniques", In: Das, S.K., Dao, TP., Perumal, T. (eds) Nature-Inspired Computing for Smart Application Design. Springer Tracts in Nature Inspired Computing. Springer, Singapore.2021. [http://dx.doi.org/10.1007/978-981-33-6195-9_1]

[27] Santosh Das, Nilanjan Dey, Ruben Gonzalez Crespo, and Enrique Herrera-Viedma, "A non-linear multi-objective technique for hybrid peer-to-peer communication", *Information Sciences,* vol. 629, pp.

413-439, 2023.
[http://dx.doi.org/10.1016/j.ins.2023.01.117]

[28] "Ashish Kumar Das and Santosh Kumar Das", *An Optimal Data Communication System in Wireless Sensor Network Using Artificial Neural Network,* pp. 7343-7346, 2021.
[http://dx.doi.org/10.4018/978-1-6684-7343-6.ch007]

[29] N. De, *Data-Driven Approach for Bio-medical and Healthcare.* Springer Nature, 2022.
[http://dx.doi.org/10.1007/978-981-19-5184-8]

CHAPTER 13

Classification of Deep Learning Techniques for Object Detection

Aras Amruth Raj Purushotham[1], Manjunath Ravindra[1,*] and Chaya Ravindra[1]

[1] *School Of CSE, REVA University, Bengaluru, India*

Abstract: The object detection framework recognises real-world objects within the frame of a moving photograph or computer-generated image. The object has a location to flow to through other objects, such as people or automobiles. Item detection is widely used in sectors where it is necessary for an organization's security and growth. The vast range of applications for protest detection include image recovery, security strategy, reason for inspection, machine framework assessment, and computerised vehicle structure. In contrast to conventional object localization techniques, machine learning-based object identification makes use of the machine's greater capacity to learn and represent knowledge [1]. A difficult problem in the analysis of designs and computer frameworks is object detection. Later on, the relationship between object detection, video analysis and image processing was developed. The complicated structure that is now being constructed includes both fundamental and sophisticated features, and the evaluation is carried out depending on the classifiers used. A complex system that can accurately assess and distinguish between numerous aspects is produced as a result of this combination. Several deep-level characteristics have been developed as a result of machine learning advancements to address the problems in the old design [2]. We conducted research on one-stage and two-stage object detectors, which are further categorised into deep learning methodologies. To enhance object detection, CNN networks employ these algorithms. An evaluation of the machine learning method for object detection is presented in this paper [3]. The protest site's applications have been distilled. The various methods of object localization employ template-based, region-based, and portion-based methods.

Keywords: Algorithms, Computer-generated image, Computer vision, CNN networks, Face recognition, LiDAR.

INTRODUCTION

A crucial topic of research in the computer vision and deep learning fields is object recognition. For increasingly complex computer vision applications, such as scene tracking and light sensing, it is a vital requirement. Other requirements include LiDAR, face recognition, event detection, behavioural analytics, semantic comprehension, and many other technologies [4]. It is intended to find and. It is

* **Corresponding author Manjunath Ravindra:** School Of CSE, REVA University, Bengaluru, India; E-mail: manjunathravindra123@gmail.com

intended to find and determine the target category of interest accurately from the image and specify each bounding box target. The focus should not only be on categorising photographs into different groups but also on properly identifying the idea and placement of the things present in each image. With the introduction of more compact, high-quality, and reasonably priced cameras as well as significant improvements in computing power and parallel processing technology, computer vision is now able to detect objects in real time by using cutting-edge hardware like multi-core processors and GPUs [5]. This has created a wide range of opportunities for businesses and people alike, enabling the development of sophisticated and effective solutions in numerous domains. Computer vision has developed into a crucial tool for navigating our increasingly data-driven environment, from self-driving cars to medical imaging [6]. We may anticipate much more sophisticated computer vision applications in the future, which will spur innovation and influence how we live and work.

The rapid development of deep convolution neural networks (CNNs) and the rise in GPU computing capacity are principally responsible for the rapid advancement in the field of computer vision-based object recognition and tracking [7].

Let's delve into the evolution of Deep Learning (DL) from Machine Learning (ML) and examine the key differences between them. The identification of patterns in data is the focus of machine learning (ML), a branch of artificial intelligence (AI). Whether the data is labelled, unlabelled, or both, a machine's ability to access and learn from it enables a variety of learning techniques, including supervised, unsupervised, semi-supervised, and reinforcement learning [8]. On the other hand, DL is a more sophisticated and precise variation of ML. Through a series of tasks, it requires the application of particular algorithms to learn both the data and the representations of the data [9]. The quantity of samples or experiences that DL is exposed to affect how well it performs. Finding and identifying items in an image is the goal of object detection. Feature extraction, informative region selection, and classification are the three steps used by conventional object detection algorithms to accomplish this goal. To locate all potential places of objects, the image is initially scanned using a sliding window at various scales. When the number of windows increases, and there are too many redundant windows generated, this process becomes computationally demanding [10]. On the other hand, if there are only a few sliding windows, some object sections might be missed. The second phase involves extracting characteristics from the chosen regions in order to represent the objects [11]. The third stage involves using a classifier to identify the kinds of objects that are present in each location. Feature extraction, which aims to discover distinctive traits of various objects in an image, is the second phase in the conventional object detection method. To extract reliable representations of objects, visual features like SIFT

(Scale Invariant Feature Transform), HOG (Histogram of Oriented Gradients), and Haar-like are used [12]. When comparing the similarity of two photos, the features that are extracted from one image are represented as vectors. However, designing reliable feature descriptors that can precisely identify items in an image can be difficult, particularly when there are differences in lighting and backdrop conditions. Because of these variances, it may be challenging to consistently extract useful features that can be utilised to recognise objects in various photos. A computer vision algorithm called SIFT (Scale-invariant feature transform) is used to find, describe, and match features in images. It was created in 1999 by David Lowe and has a variety of uses, including object recognition, robotic mapping and navigation, image stitching, 3D modelling, gesture recognition, video tracking, and animal identification by a human. In order for the process to function, a set of reference images must first have SIFT key points extracted from them and stored in a database. In order to find probable matching features based on the Euclidean distance between their feature vectors, each feature in a newly processed image is compared to the database. The computer then chooses subsets of important points that, after weeding out false matches, depict the object in the new image together with its location, size, and orientation. To accomplish this, a hash table version of the generalised Hough transform is used to efficiently group consistent matches [13]. The method then determines the probability that a given set of features indicates the existence of an object, taking into account the precision of the fit and the potential number of false matches. It is quite certain that object matches that pass this set of tests are accurate. A feature descriptor known as HOG (Histogram of Oriented Gradients) is frequently employed in computer vision and image processing for object recognition [14]. It operates by counting the occurrences of a gradient orientation in particular regions of a picture. There are some significant discrepancies between this approach and edge orientation histograms, SIFT descriptors, and shape contexts. A dense grid of cells with equal spacing is used to calculate HOG, and overlapping local contrast normalisation is used to increase accuracy. In the HOG procedure, the image is converted into a feature vector with a user-specified length. Although the final image might be difficult to read, it is suitable for use in support vector machines (SVMs), an image classification approach that can yield precise results. An instrument used to distinguish a target object from the total image is a classifier. Making the representations more structured, meaningful, and helpful for visual recognition, it serves to distinguish the target object from other categories [15]. A more hierarchical, semantic, and educational approach to visual identification is made possible by the use of classifiers. In the realm of computer vision and image processing, Support Vector Machines (SVM), AdaBoost, and Deformable Part-based Models are three of the most popular classifiers (DPM) [16]. DPM is regarded as the most adaptable of these as it accounts for object deformations through its deformation cost. The development of extremely accurate part-based models for a range of object classes is made possible by the use of graphical models in discriminative learning. Support Vector Machines (SVM) is a classification-focused machine learning model. In order to establish as wide a distinction between the categories as feasible, it works by mapping samples from various categories to points in space. SVMs convert inputs into high-dimensional feature spaces, which they then use to perform linear and non-linear classification. An SVM training method develops a model that can classify incoming examples into one of two categories based on a series of training examples that have been labelled as belonging to one of the categories. As a result, SVM qualifies as a non-probabilistic binary linear classifier that can distinguish between two categories using a linear boundary. In the field of computer vision, deep learning models are frequently employed for both general and domain-specific object detection and tracking. Convolutional Neural Network (CNN) models are used to extract information from the input frame for object detection [17]. The items in the frame are then classified using these extracted features. Based on the outcomes of the object detection, the objects are tracked from frame to frame in object tracking. To establish the item's closeness between frames, the attributes of the object are compared, which enables precise tracking of the object across time. Deep Learning models have been used successfully to improve accuracy and efficiency in the detection and tracking of objects. Finding and identifying items in a picture or video is what object detection entails. Deep Learning models or image processing methods can be used to do this. Computer vision and algorithms are used in image processing, whereas Deep Learning models, such CNNs, learn to detect things. The requirements and difficulties of the task determine the strategy to choose. Unsupervised image processing is affected by a variety of variables, including lighting conditions. Contrarily, DL-based object detection requires supervision, GPU computational capacity, and a sizable training dataset. Several well-known datasets are Microsoft-produced massive object detection and segmentation effort known as MS COCO (Common Objects in Context). It is intended to identify and separate things within a picture, giving users a thorough comprehension of the context and connections between objects. For those working in the fields of computer vision and image processing, MS COCO offers an invaluable resource. It entails identifying and locating both 2D and 3D objects. The characteristics of the object attributes define the interaction between objects. The dataset is composed of 80 COCO classes for objects, such as people, chairs, buses, and other easily recognisable objects. This dataset has problems with high-quality graphics or frames because it was made to advance image recognition. The KITTI dataset includes 7481 training photos with 3D bounding boxes, which are typically used in platforms for autonomous driving. It includes a variety of tasks, including stereo, optical flow, visual odometry, *etc.* ImageNet is a sizable image collection that is organised according to the WordNet hierarchy, which only contains nouns. Each hierarchy node contains hundreds or thousands of photos. Researchers and professionals working in the fields of computer vision and image processing are free to use the data in ImageNet. It is a crucial tool in the area since it offers a wealth of resources for jobs involving image categorization and recognition. Generic object detection is one of the various approaches to object detection. Identifying and locating objects in a picture, as well as labelling them with bounding boxes to indicate their confidence level, are the goals of generic object detection. Two different types of detectors—two-stage and one-stage—are used to accomplish this. The one-stage strategy predicts all bounding boxes in a single pass through a neural network, whereas the two-stage approach uses a typical object detection process that includes object localization and classification.

Literature Survey

Majumder *et al.* suggested that in every organization, effective data collection, classification, and oversight are crucial components. To accomplish this, enterprises invest significant amounts of resources, both in terms of manpower and finances, to ensure that this information is properly managed and structured. This process is commonly referred to as Knowledge Management (KM) [1]. Das *et al.* proposed a statement that over the past few decades, smart applications have been developed to meet the growing needs of users. These applications are designed to tackle real-world problems and are based on specific user requirements and specifications. To address the diverse and complex nature of these requirements, intelligent techniques such as machine learning, nature-inspired computing, soft computing, and computational intelligence are used under the umbrella of artificial intelligence. The techniques are tailored to the specific parameters of each problem, which can range from precise to imprecise. This chapter provides an overview of smart design and its applications, highlighting the challenges and techniques associated with their development [2]. Dey *et al.*, in their book, presented the latest research in machine learning, artificial intelligence, and data analytics for biomedical and healthcare applications. It explores the difficulties involved in analysing biomedical data, such as the complexities associated with higher dimensions, class imbalances, and smaller database sizes. The book also focuses on the development of new pattern recognition and machine learning techniques that are specifically designed for medical and genomic data, which is a critical but challenging task. It is an essential resource for healthcare professionals who have access to valuable data sources but require additional expertise in data mining to use them effectively [3]. Xie *et al.* proposed that remote sensing researchers are increasingly focusing on oriented object detection because it offers a more precise spatial representation of oriented target objects [4]. Despite significant progress in this area, detecting objects with multiscale and arbitrary directions remains a challenging task. To address this issue, this paper proposes a multilevel stacked context network (MSCNet) that leverages semantic relationships between different objects and contexts in remote sensing images to enhance target detection accuracy. The paper also explores the feasibility of using a Gaussian distribution to represent objects, which can help to overcome the limitations of traditional oriented bounding box representation. Experimental results on two common remote sensing datasets demonstrate that the proposed MSCNet outperforms the baseline approach [5]. Ren *et al.* presented a new method for object detection using a Region Proposal Network (RPN) that shares convolutional features with the detection network, which reduces computation time and improves accuracy. The RPN predicts object boundaries and objectness scores at each position and is trained to generate high-quality region proposals that are used by the detection network for object detection. Our system achieves state-of-the-art accuracy on the PASCAL VOC 2007 and 2012 datasets with a frame rate of 5fps on a GPU [6]. We plan to release the code for our method.

Two-stage Detectors

Two-stage detectors have two tasks: object region proposal and object categorization. The first step in the object detection process is to propose an object region using conventional or deep learning methods. Afterwards, the region is used to extract features for classification, increasing detection precision. The most often used two-stage detectors are RCNN, Fast RCNN, Faster RCNN, Mask RCNN, RFCN, FPN, Granulated CNN, and Granulated RCNN, and their fundamental architecture is described in more depth below [6].

One-stage Detectors

Although these detectors are quicker than two-stage detectors, their accuracy may be worse. One-stage detectors consider detections to be regression by learning from the input image and classifying the data by speculating on potential bounding boxes. YOLO, DetectNet, SSD, and SqueezeDet are a few examples.

Deep Learning Models

Let's first examine the fundamental architecture of CNN before looking at object detection utilising deep learning techniques.

Convolutional Neural Networks (CNN), Artificial Neural Networks (ANN), and Recurrent Neural Networks (RNN) are a few deep learning models, and CNN has made a significant contribution to the field of computer vision.

CNN

Deep neural networks, including convolutional neural networks, are frequently employed in the fields of computer vision and pattern recognition.

They can recognise and distinguish between different aspects of an image, then use those features to forecast outcomes.

CNNs consist of multiple layers, including convolutional layers, pooling layers, and fully connected layers. Convolutional layers apply a set of filters to the input image, producing a set of feature maps that highlight different aspects of the image. Pooling layers downsample the feature maps, reducing their size and making the network more efficient.

The output from the convolution process is used by the fully connected layer of the network to provide predictions based on the features extracted in the earlier stage.

During training, the CNN learns the optimal values for the filter weights and biases by adjusting them to minimize the error between the predicted output and the actual output. This process is typically done using backpropagation, where the error is propagated back through the network to adjust the weights and biases.

CNNs have achieved state-of-the-art performance on many image recognition tasks, including object recognition, face recognition, and image segmentation. They have also been used in a variety of other applications, such as natural language processing and speech recognition.

Convolutional Layer

The features from the input images are extracted in the initial layer of convolutional neural networks at this point. The convolution function is applied to the input image and the target image, and the feature from the input image is removed by sliding the target image over the input image and taking the product of the coordinates. The size of the matrix is used to apply every mathematical operation (MxM) [18].

The following layer receives this feature-extracted matrix and processes it further.

Pooling Layer

The Pooling Layer follows the Convolutional Filter, and its objective is to reduce the extracted feature map's size in order to reduce computation. Either maximum pooling or average pooling can be used to accomplish this. The Pooling Layer reduces the connections between layers by taking the maximum or average value from a set of values when applied to individual feature maps. It is normally positioned between the Convolutional Layer and the Fully Connected Layer [19].

Fully Connected Layer

The fully connected layer is a layer of a network that connects earlier layers and processes incoming data using neurons. In order to extract useful information, this layer flattens the output of the preceding layer into a vector and performs a series of mathematical operations. A Convolutional Neural Network architecture's final layer receives its input from this layer's output.

Dropout Layer

Since every flattened vector feature is connected to a fully connected layer, the training dataset lead to overfitting. Overfitting can happen when a model performs flawlessly on a training dataset but fails to perform flawlessly when additional raw data is fed into the model. Thus, the Dropout Layer is used to reduce the neural network's size by removing part of its neurons.

Activation Functions

The most crucial layer for any CNN model is this one. They determine whether the network should continue or end and are used to understand complex network variables. ReLU, tanH, Softmax, and sigmoid are just a few of the activation methods that can be used. Both sigmoid and Softmax functions are employed in the binary classification model. The most prevalent activation function, ReLU, eliminates negative values and keeps output values between 0 and infinity. ReLU is zero if the value is negative.

Generic Object Detection

As previously mentioned, two-stage and one-stage detection approaches can be used to categorise generic object detection techniques for detection that employ a two-stage process. Here are explanations of algorithms:

RCNN

When more than one object appears in an image on CNN, CNN can be applied to a region using the sliding window brute force method, although the image may contain the same object in multiple sizes. As a result, there are more regions that require more computing and hence are more expensive [20].

The goal of R-CNN is to precisely identify the target object in an input image by creating a bounding box around it. R-CNN seeks to find an object in an image to address a problem. There are three components to this system. First, the region-unaffected system category. These specify a collection of an image's coordinates. The second method is the convolutional network, which extracts features from a region. The third is a classifier that includes a collection of classifications, like SVM.

R-CNN has several advantages over earlier object detection methods. First, it achieves state-of-the-art accuracy on standard object detection benchmarks. Second, it can detect objects of different sizes and aspect ratios, which makes it more versatile than earlier methods. Third, it can be fine-tuned for specific object classes or domains, which makes it adaptable to a wide range of applications.

However, R-CNN also has some limitations. The region proposal stage is computationally expensive, which makes it slow to train and test. Additionally, the selective search algorithm used for region proposals is not very accurate, which can lead to missed detections or false positives. To address these issues, subsequent models like Fast R-CNN and Faster R-CNN were developed, which use more efficient region proposal methods and end-to-end training techniques to improve accuracy and speed. Models like Fast R-CNN and Faster F-CNN will be further discussed.

Ross Girshick and his colleagues developed a method that uses selective search to extract only 2000 regions from an image in order to overcome the difficulty of needing to evaluate a huge number of regions. The term "region proposals" is used to describe these areas. By putting a cap on the number of regions at 2000, the emphasis is shifted from trying to categorise many areas to just assessing these 2000 region proposals. These 2000 region proposals are produced using the selective search algorithm, which is covered in more detail below.

The system processes 2000 candidate region proposals, which are squared and fed into the convolutional neural network (CNN) as input. A 4096-dimensional feature vector is the output of the CNN's function as a feature extractor. An SVM is then used to classify the presence of an object within the suggested candidate region using the dense layer of information from the image. The system not only anticipates the presence of an object but also forecasts four offset values to improve the bounding box accuracy. For instance, the algorithm might foretell the presence of a person in a given area, but it also considers the distinctive face traits present in that area to increase the accuracy of its forecasts.

Selective Search Algorithm

1. The system creates many alternative regions for an initial sub-segmentation.

2. Through the use of a greedy algorithm that repeats itself, comparable sections are combined with larger ones.

3. The regions created during the earlier phases are used to develop the final candidate region suggestions.

Feature extraction from Region Proposals.

At this point, convolutional neural networks generate 4096-dimensional feature vectors from 2000 regions for each image (CNN).

1. There are 5 convolutional layers and 2 fully connected layers in the AlexNet defined by Krizhevsky *et al.* 1000 photos are used in the initial training of CNN to extract image attributes.

2. Regardless of the original size, all 2000 regions from the image obtained using the selective search technique are reduced to a predetermined size of 227 × 227 before being sent to CNN.

3. Each object that has the greatest IoU overlap is given a positive label. Boxes are negative and considered background class if Iou is 0.5.

4. For each image's 2000 regions, a 4096-dimensional feature vector is created after the classification layer.

SVM for object classification:

In this stage, a linear SVM classifier is used to detect the presence of objects belonging to a specific class for each class or region.

1. Classifier takes an input of 4096 feature vectors for each region.

2. Positive features are those that come from a bounding box. If the Intersection over Union (IoU) of two bounding boxes is less than 0.3, it is regarded as a negative match in the context of object detection. On the other hand, the region is ignored when training the Support Vector Machine (SVM) classifier if the IoU overlap is more than 0.3 yet does not match a bounding box. This criterion aids in ensuring that only accurate and pertinent information is used during the training process, improving performance in object detection.

3. The test time for a single image region is calculated using the dot product of the feature matrix and the SVM weights. A feature matrix of 2000 regions by 4096 is produced for an input image. As a result, the 4096 x N SVM weight matrix is due to the N classes.

4. SVM output is a result of positive objects with classes over all 2000 regions after training.

Problems with RCNN

1. It can take a while to categorise 2000 region recommendations for each image in order to train the object detection network. The network's training process is made slower and more difficult by the numerous region proposals per image.

2. Real-time implementation is difficult to achieve because the testing process for each image takes about 47 seconds.

3. The chosen search algorithm is a fixed algorithm. Hence, no learning occurs at that time. This can lead to numerous proposals for terrible candidate regions.

Fast RCNN

To create a feature map, the input image is processed through a number of convolutional and pooling layers. The pooling layer is then used to determine object regions of interest (RoI), after which a fixed-length feature vector is retrieved for each RoI from the feature map. A fully connected layer that connects to the output layer processes the feature vector and includes a branch for category-specific bounding box regression and a softmax branch for categorization.

Fast R-CNN is a faster and more efficient version of R-CNN that was introduced by Ross Girshick in 2015. Like R-CNN, Fast R-CNN is an object detection model that uses convolutional neural networks (CNNs) to classify and localize objects within an image. However, it improves R-CNN by addressing some of its limitations, particularly in terms of speed and memory usage.

Another important feature of Fast R-CNN is its use of the "RoIAlign" layer, which addresses the misalignment problem that arises from the discretization of the RoI pooling layer used in R-CNN.

In addition to these technical improvements, Fast R-CNN has also been used for a variety of practical applications, such as object detection in images and videos, scene recognition, and instance segmentation. It has been shown to achieve state-of-the-art results on several benchmark datasets, and its efficient implementation makes it a popular choice for real-time object detection tasks.

Fast RCNN has a faster processing time than RCNN because of this. Fast RCNN does not feed 2000 regions into a convolutional neural network, in contrast to RCNN. Instead, each image undergoes a single convolutional operation to produce a feature map with the features of several regions of various sizes. In comparison to RCNN, this simplified technique produces a faster and more effective object detection process. The truncated single value decomposition (SVD) approach, which provides matrices with a defined number of columns, is used to find weight parameters more quickly and accelerates execution by compressing the fully linked layer. The methods used in their training are one of the main distinctions between RCNN and Fast RCNN. Fast RCNN uses a simplified, single-stage training strategy, while RCNN uses a multi-layered, multi-stage approach. Fast RCNN trains more quickly and effectively than RCNN due to the difference in the number of stages and layers. Fast RCNN uses the selective search approach since the pooling map proposed by the convolutional feature map slows down the operation. The Fast RCNN increases effectiveness and accuracy while saving money on storage space costs.

Polling Region of Interest (RoI): In the Fast RCNN, RoI polling was developed. It is used to produce feature maps of the same fixed size from a variety of input images that contain RoI.

Region of Interest (RoI) takes 2 inputs:

1. Convolution neural network layer's extracted features map

2. The first two coordinates (r,c,h,w) represent the region of interest's upper and left, the following two are its height and width, and N is the number of RoIs.

Faster RCNN

Faster RCNN comprises of two primary components:

1. the Region Proposal Network (RPN).

2. Object Detection Network.

The foundation of the Faster RCNN design is the CNN architecture. Both Region Proposal and Object Detection Network include CNN.

CNN Layer

In trials, 5 and 13 convolutional layers in ZF and VGC-16 networks are used, respectively [21]. Before entering the area proposal network, a stride of 16 is used to shrink the feature map from its initial size of 1000 x 600 to 62 x 37.

The Region Proposal Network (RPN):

Taking input of convolution layer feature map from backbone layer and generating Anchors by Sliding Window on Input Feature Map

Anchors

For each sliding window applied to the input feature map by the Region Proposal Network (RPN), k numbers of anchor boxes are generated. As a result, a convolutional feature map of size w x h has anchor boxes of size N = w x h.

The object proposed is transferred to the intermediate convolutional layer 3*3 and one padding layer, and an output of 256 for ZF and 512 for VGG-16 is observed.

Object Localization and Classification System

Faster RCNN uses a similar method as Fast RCNN for object detection. In this technique, fixed-size area suggestions are created using a RoI pooling layer and evaluated by a softmax classifier and regressor bounding box. By this method, objects may be found and matching bounding boxes can be created.

RoI Pooling

The RoI pooling layer, which serves the same function as in Fast R-CNN, receives the output from the region proposal. The RoI pooling layer's job is to turn the variable-sized region suggestions produced by RPN into a fixed-size feature map. This article has covered the RoI pooling layer in great detail. The RoI pooling layer produces an output with a size of (7 x 7 x D), where D is equal to 256 for ZF and 512 for VGG-16.

The feature map created from the RoI pooling layer is used as input in the final stage of object detection in Faster RCNN, where it is then sent *via* a fully connected layer. The resulting feature matrix is then flattened and passed to the Softmax and Bounding Box Regression Layers, which are two more fully connected layers. The Softmax layer, which has N+1 output parameters, is the first layer in the object detection process [22]. Predictions about the presence of objects in the region proposal are made by this layer. The placement of items within the image is determined by the second layer, a regression layer for bounding boxes with 4 x N outputs.

Compared to Fast RCNN, which relies on selective search for region proposal generation, Faster RCNN has a built-in Region Proposal Network that improves the performance of the region proposal model during training. As a result, the overall detection time is decreased. An improvement in mean Average Precision is seen after the Region Proposal Network (RPN) with selective search is added (mAP).

ONE-STAGE OBJECT DETECTORS

A one-stage detector only needs to make one run through the neural network and predict all of the bounding boxes at once. This is substantially faster and much more suited to portable devices. One-stage detectors are most frequently used in YOLO (You Only Look Once) and SSD technologies (Single Shot Detector).

YOLO

YOLO is a real-time neural network-based object detection system. It has been utilised in a range of applications, including the detection of traffic signs, persons, and parking metres [14]. It is renowned for its speed and accuracy. The object detection process is carried out by YOLO as a regression problem, and the class probabilities of the observed items are output. It uses three main strategies and is notable for its speed, accuracy, and learning capabilities.

1. Residual units.

2. Intersection Over Union Regression for Bounding Boxes.

3. Intersection Over Union (IOU).

Residual Blocks

The image is initially split up into smaller grid cells, each measuring S x S. Each grid cell is in charge of identifying items that show up there. For instance, if the centre of an object is visible in a certain grid cell, then that cell is in charge of identifying it [15]. By breaking the image up into grid cells, YOLO is able to break the process of object identification and recognition into more manageable, smaller chunks.

Bounding Box Regression

In order to determine the height, width, centre, and class of the objects in the image, YOLO uses a single bounding box regression technique. With this method, YOLO can quickly and precisely recognise and categorise objects in an image.

Intersection Over Union (IOU)

An important component of object detection is the Intersection Over Union (IOU), which assesses how much the predicted bounding boxes and the real objects overlap. In order to create a bounding box that exactly encloses the items, YOLO uses IOU. In YOLO, the bounding boxes and their confidence scores are predicted for each grid cell. The projected bounding box must exactly match the real bounding box in order for the IOU to be equal to 1, removing any other bounding boxes [16]. By doing this, it is made possible to guarantee that the anticipated bounding boxes appropriately depict the objects' dimensions, such as their height and width. Without any extra boxes that do not fit the object's requirements, the final detection result consists of singular bounding boxes that properly fit the objects. The below image provides an illustration of how IOU operates.

SSD (Single Shot Multibox Detector)

The SSD (Single Shot MultiBox Detector) is a powerful object detection method that is quicker than Faster RCNN. Its two pieces are an SSD head and a backbone network. The feature extractor of the backbone network is frequently an image classification model that has already been trained. The backbone network is often ResNet, which has been trained on ImageNet. The final fully connected classification layer is deleted, leaving a deep neural network that can extract semantic meaning from an input image while preserving the spatial structure of the image at a lower resolution. For instance, when ResNet34 is employed as the backbone network, 256 7x7 feature maps are generated for each input image [23].

One or more convolutional layers are placed into the backbone network as the SSD head. The outputs of the SSD head represent the bounding boxes and object classes that are present in the spatial region of the final layer activations.

CONCLUSION

The YOLO technique is presented in this paper with the aim of using a single neural network for object detection. This algorithm works better than others since it is broader. Many diverse strategies are used in many domains when generalising from natural photos. The suggested algorithm is easy to create and maintain, and it may be trained on a whole image or easily customised for a particular area. Using methods, the classifier is constrained to a particular area. When YOLO anticipates borders, it can use the entire image. Also, it anticipates a decrease in false positives in the background areas. YOLO method is superior to other classification algorithms when compared, and it is also noticeably quicker and more effective to use in practical settings.

REFERENCES

[1] S. Majumder, and N. Dey, *AI-empowered knowledge management.* Springer, 2022. [http://dx.doi.org/10.1007/978-981-19-0316-8]

[2] S.K. Das, *Smart design and its applications: Challenges and techniques.* Nature-Inspired Computing for Smart Application Design, 2021, pp. 1-6.

[3] S.K. Das, N. Dey, R. González Crespo, and E. Herrera-Viedma, "A non-linear multi-objective technique for hybrid peer-to-peer communication", *Inf. Sci.,* vol. 629, pp. 413-439, 2023. [http://dx.doi.org/10.1016/j.ins.2023.01.117]

[4] A.K. Dass, and S.K. Das, "An optimal data communication system in wireless sensor network using artificial neural network", In: *Constraint Decision-Making Systems in Engineering* IGI Global, 2023, pp. 120-134. [http://dx.doi.org/10.4018/978-1-6684-7343-6.ch007]

[5] N. Dey, Ed., *Data-Driven Approach for Bio-medical and Healthcare.* Springer Nature, 2022.

[6] X. Xie, "Oriented R-CNN for object detection", *Proceedings of the IEEE/CVF International Conference on Computer Vision,* 2021.

[7] S. Ren, "Faster r-cnn: Towards real-time object detection with region proposal networks", *Adv. Neural Inf. Process. Syst.,* p. 28, 2015.

[8] W. Zhiqiang, and L. Jun, "A review of object detection based on convolutional neural network", *2017 36th Chinese control conference (CCC),* 2017. [http://dx.doi.org/10.23919/ChiCC.2017.8029130]

[9] A. Dhillon, and G.K. Verma, "Convolutional neural network: A review of models, methodologies and applications to object detection", *Progress in Artificial Intelligence,* vol. 9, no. 2, pp. 85-112, 2020. [http://dx.doi.org/10.1007/s13748-019-00203-0]

[10] C. Eggert, *"A closer look: Small object detection in faster R-CNN." 2017 IEEE international conference on multimedia and expo (ICME).* IEEE, 2017.

[11] J. Tompson, "Efficient object localization using convolutional networks", *Proceedings of the IEEE conference on computer vision and pattern recognition,* 2015.

[12] H. Harzallah, F. Jurie, and C. Schmid, "Combining efficient object localization and image classification", *2009 IEEE 12th international conference on computer vision,* 2009. [http://dx.doi.org/10.1109/ICCV.2009.5459257]

[13] P. Jiang, D. Ergu, F. Liu, Y. Cai, and B. Ma, "A review of yolo algorithm developments", *Procedia Comput. Sci.,* vol. 199, pp. 1066-1073, 2022. [http://dx.doi.org/10.1016/j.procs.2022.01.135]

[14] W. Fang, L. Wang, and P. Ren, "Tinier-YOLO: A real-time object detection method for constrained environments", *IEEE Access,* vol. 8, pp. 1935-1944, 2020. [http://dx.doi.org/10.1109/ACCESS.2019.2961959]

[15] Juan Du, "Understanding of object detection based on CNN family and YOLO", Journal of Physics: Conference Series. Vol. 1004. IOP Publishing, 2018. [http://dx.doi.org/10.1088/1742-6596/1004/1/012029]

[16] C. Liu, "Object detection based on YOLO network", *2018 IEEE 4th information technology and mechatronics engineering conference (ITOEC),* 2018. [http://dx.doi.org/10.1109/ITOEC.2018.8740604]

[17] Z. Zou, "Object detection in 20 years: A survey", *Proceedings of the IEEE,* 2023. [http://dx.doi.org/10.1109/JPROC.2023.3238524]

[18] Yali Amit, Pedro Felzenszwalb, and Ross Girshick, "Object detection", *Computer Vision: A Reference Guide,* pp. 1-9, 2020. [http://dx.doi.org/10.1007/978-3-030-03243-2_660-1]

[19] M. Tan, R. Pang, and Q.V. Le, "Efficientdet: Scalable and efficient object detection", *Proceedings of the IEEE/CVF conference on computer vision and pattern recognition,* 2020. [http://dx.doi.org/10.1109/CVPR42600.2020.01079]

[20] C. Szegedy, A. Toshev, and D. Erhan, "Deep neural networks for object detection", *Adv. Neural Inf. Process. Syst.,* p. 26, 2013.

[21] L. Alzubaidi, J. Zhang, A.J. Humaidi, A. Al-Dujaili, Y. Duan, O. Al-Shamma, J. Santamaría, M.A. Fadhel, M. Al-Amidie, and L. Farhan, "Review of deep learning: Concepts, CNN architectures, challenges, applications, future directions", *J. Big Data,* vol. 8, no. 1, p. 53, 2021. [http://dx.doi.org/10.1186/s40537-021-00444-8] [PMID: 33816053]

[22] R. Girshick, "Rich feature hierarchies for accurate object detection and semantic segmentation", *Proceedings of the IEEE conference on computer vision and pattern recognition,* 2014. [http://dx.doi.org/10.1109/CVPR.2014.81]

[23] W. Liu, "Ssd: Single shot multibox detector", *14th European Conference.* Amsterdam, The Netherlands, pp 21-37, 2016.

SUBJECT INDEX

A

Acid, ethylenediaminetetraacetic 4
AI 5, 127
  -empowered knowledge management 127
  -powered tools and machine learning 5
Air 34, 35,130
  polluted 35
  pollution 34, 35
Algorithms 10, 11, 12, 13, 44, 45, 73, 156, 157, 158, 173, 177, 178, 212, 213, 214, 219, 220
  computational 73
  computer-based 73
  computer vision 214
  garbage collection 173
  meta-heuristic 12
Analysis 121, 162
  automatic 162
  bibliometric 121
ANSYS software 35
ANT colony optimization 13
Artificial 1, 12, 13, 14, 26, 47, 48, 124, 153, 158, 163, 164, 193, 217
  bee colony (ABC) 12
  erythrocyte 1
  immune systems (AIS) 14
  intelligence techniques 26, 48
  neural networks (ANN) 13, 47, 124, 153, 158, 163, 164, 193, 217
Automated 84, 116, 127
  data processing 84
  services 127
  vcrification systcm 116
AVIPSA tool 121

B

Bacteria-based 4
  nanorobotic system 4
  nanorobotics 4
Blockchain 85, 86, 89, 94, 95, 98, 99, 100, 101, 103, 104, 105, 106, 107, 108, 114, 118, 122, 123, 124, 125, 126, 131, 133
  consortium 101, 118, 133
  hybrid 101
  -integrated IoT supply chain system 131
  integration 89, 124
  systems 106
  technology applications 95, 123
Blockchain applications 95, 98, 100, 127
  healthcare-based 100
Blockchain-based 94, 95, 102
  healthcare applications 102
  identity management systems 94
  payment systems 94, 95
Blockchain network 100, 101, 109, 118, 119
  hybrid 101
Blockchain technology 94, 108, 123
  adoption of 94, 108
  framing 123
Blood pressure 122
BMRFO technique 45

C

CAD software 39
Cloud 64, 65, 66, 69, 73, 84, 86, 87, 104, 105, 106, 128
  climate 106
  computing 64, 66, 69, 86, 87, 128
  encryption 104
CNN 212, 223
  architecture 223
  networks 212
Coal-fired power plant 190
Cognitive disorders 140
Communication systems 73
Comprehensive system adaption 67
Computer vision 212, 213, 214, 215, 216, 217, 218
  and algorithms 215
Constraint application protocol 129

Control 4, 17, 59, 73, 98, 106, 114, 115, 128, 131, 132, 192
automatic 131
environmental parameter 114
power system operation 59
Convolutional neural network (CNN) 153, 156, 213, 215, 217, 218, 219, 220, 221, 222, 223
Cryptocurrency mining 108
Cryptographic angles 104
Cryptography 124
Cyber-physical system (CPS) 67, 68, 70, 72, 73, 74
Cybersecurity and cyber-physical systems 70

D

Data 47, 83, 84, 93, 95, 128, 216
mining 47, 216
processing tools 128
science and artificial intelligence methodologies (DSAIM) 83, 84, 93, 95
Deep 14, 18, 21, 218, 226
neural networks 18, 21, 218, 226
Q-learning algorithm 14
Deep learning 9, 11, 14, 48, 153, 154, 162, 163, 165, 213, 215, 217
algorithms 162
methods 163, 217
Devices 1, 3, 4, 5, 34, 38, 84, 115, 116, 117, 126, 127, 128, 130, 171, 176
artificial 1
biomedical 5
electronic 171
mobile 84
nanorobotic 3
portable electronic 176
Digital assets and cryptocurrencies 94
Discovery, biomedical 106
Discrete wavelet changes 104
DNA Nanobots 3
Dopamine 139, 140, 144, 145, 146, 147, 149
and serotonin 144, 146
function 149
low 140
pathway 139
surges 139
systems 145
Dragonchain technology 101
Drug discovery processes 121
DSAIM techniques 88
DSMC method 6
DSSA technique 45
Dynamic(s) 70, 105, 129
association 105
RAM 129
simulate system 70

E

Earthquake 156, 158, 159
damage 156
database 158
forecasts 158
prediction method 158
seismograph 159
Effective marketing tactics 89
Effects of Earthquakes 164
Electric 59, 189, 190, 191, 192, 193
power system 189
vehicles (EVs) 59, 189, 190, 191, 192, 193
Electrical power systems 44
Electricity 44, 66, 189, 190
consumption 189, 190
Electronic(s) 98, 103, 104, 105, 121, 174
flash-based consumer 174
health records (EHRs) 98, 103, 104, 105, 121
Elements 128, 168
complicated mechanical 168
mechanical 128
Employee 136, 137, 138, 142, 143, 149
attrition 136, 138
retention 138, 142, 143, 149
turnover 136, 137
Endurance, extended NAND flash memory 176
Energy 4, 5, 35, 44, 168, 171, 189, 190, 192
consumption 171
converting chemical 4
economy 168
harvesting 5
kinetic 4
renewable 35
Environment 2, 9, 10, 11, 12, 13, 14, 15, 19, 20, 26, 30, 44, 68, 72, 73, 98
digitalized medical care 98
productive industrial 68
Ethereum-based smart contract method 107

F

Factories, frozen food 126
Fatal 35, 138
accidents 138
disorders 35
Financial 83, 86, 87, 88
service customers 87
technology 83, 86, 88
Finite element method 34, 35
Fintech 84, 87, 88, 89, 96
companies 84, 96
ecosystem 87, 88
technologies 89
FinTech business 88, 89
areas 88, 89
challenges 88, 89
Firefly algorithm (FA) 12
Flash 168, 169, 170, 171, 172, 174, 175, 176, 178, 180, 181, 182
memory 168, 169, 170, 171, 172, 174, 175, 176, 178, 180, 181, 182
translation layer (FTL) 168, 169, 171, 172
Force 4, 40
acoustic radiation 4
actuating 4
Forecasting 13, 46, 163
weather 163
Fraud detection 13, 86, 89, 90, 95
Free blocks, managing 175
FTL 169, 171, 181
algorithm 171, 181
method 169, 181
system software 169
technique in embedded applications 169
Fuzzy logic approach 12

G

Gaussian distribution 217
GPU 213, 215
computational capacity 215
computing capacity 213

H

Hard disk drives (HDDs) 168, 169, 170
Health records, electronic 98, 103, 121
Healthcare 98, 99, 100, 103, 105, 106, 107, 114, 115, 118, 121, 122, 123, 124, 125, 126
-based privacy adoption 99
blockchain 106
data 124
industry 105, 124
landscape 122
sectors 115, 121, 122, 123, 124, 125
service providers 105
Healthcare systems 105, 122
technology-assisted 122
Heart syndromes 35
Hormonal imbalances 140
Hormone melatonin 146
Human resource planning 143
Hybrid 50, 192
chaotic slime mould technique 50
optimization algorithm 192

I

Industrial 44, 65, 66, 67, 68, 69, 73, 116, 120, 189
automation 120
businesses 65
internet of things (IIoT) 67, 68, 69, 73
mechanics 116
revolution 65, 66, 67, 73
sectors 44, 189
Industries 9, 10, 64, 66, 67, 68, 74, 75, 76, 77, 83, 86, 87, 93, 94, 105, 108, 120, 122, 126, 127, 209
autonomous 9
contemporary power 209
conventional banking 87
financial 86, 87, 93, 94
iron 66
mechanical 76
medical care 105, 108
Information 98, 108, 142
and communication technologies 98
technology division 142
transmission 108
Innovation, blockchain-based medical service 102
Insurance 87, 88, 89, 95, 104
agency 104
fintech companies 87
Integrated business processes 68

Integrating implicit motives 142
Internet of medical things (IoMT) 98, 102, 103, 108, 109, 121
Intrusion detection system (IDS) 127, 157
IoT 71, 98, 99, 115, 117, 124, 131, 133
  healthcare industry 99
  healthcare security 98
  monitoring sensors 115
  sensors 71, 117, 133
  sensors and Blockchain 131
  technologies in healthcare 124
IWOA method 192

L

LDR sensor 131
Lightweight cryptography 5
Long short-term memory (LSTM) 91, 156, 191
LSTM networks 91

M

Machine failure 116
Machine learning 47, 48, 65, 69, 86, 90, 91, 92, 93, 153, 156, 157, 159, 161, 162, 163, 212, 213, 216
  advancements 212
  algorithms 65, 90, 91, 92, 93, 161
  applications 161
  method 212
  techniques 47, 91, 92, 216
Manual monitoring of machinery 116
Markov 16, 17
  decision process (MDP) 16, 17
  process 17
Mechanisms, cryptography-related security 26
Medical service 98, 102
  industry applications 102
  information 98
Memory devices 175
Methods 48, 212
  metaheuristic 48
  portion-based 212
Mobile 9, 13, 34, 84, 89
  charger 34
  phones 84
  robots 9, 13
Motor functions 140

N

NAND memory 169
Neural networks 11, 13
  biological 13
  contemporary 11
Nontoxic urea 4
Numerical methodologies 6
NWOA method 46

O

OAGO method 46
Object detection 216, 217, 222, 223, 224, 225, 226
  method 226
  network 222, 223
  process 216, 217, 224, 225
Operations 64, 69, 73, 74, 75, 76, 168, 169, 171, 173, 197, 199, 218, 219
  autonomous 74
  mathematical 218, 219
OT equipment 70
Oxytosin 149

P

PHR designs 104
Physical health records (PHRs) 104
Plug-in electric vehicles (PEVs) 189, 190, 191, 192, 194, 197
Power 44, 59, 94, 157, 168, 194, 197
  bitcoin 94
  consumption 157, 168
  electrical 44
  industry 44, 194
  solar 59, 197
  system network 44
Power sector 43, 44, 50
  electrical 44
Practical byzantine fault tolerance (PBFT) 107
Process, mining 107
Psychological disorders 149
Psychology 138
  cognitive 138
  social 138
Public blockchain platform 100

R

Radio-frequency identification (RFID) 114, 124, 129
RAM 169
  capacity constraints 169
  memory 169
Real-time 65, 68, 222
  data management 68
  implementation 222
  visibility 65
Rect and Kovan test networks 107
Recurrent neural networks (RNNs) 13, 91, 92, 153, 217
Region proposal network (RPN) 217, 223, 224, 225
Regulation technology 89
Reinforcement learning 9, 11, 13, 14, 15, 16, 17, 160, 161, 213
  algorithms 11, 14
  approach 14
REM sleep 146
RES methodology 48
RFID sensors 131
Robot 3, 4, 9, 10, 11, 12, 13, 14, 15, 19, 20, 22, 23, 25, 26, 30
  autonomous 10, 14, 15
  multiple 10
  nucleic acid 4
Robotic(s) 3, 4, 5, 73, 93
  process automation in financial services 93
  autonomous 73
  process automation (RPA) 93
  systems 3, 4, 5
Rod-shaped nanobots 3
RPA automation tools 93

S

Schizophrenia 147
SChoA method 46
Secure financial system 94
Security 85, 86, 87, 95, 96, 103, 105, 117, 118, 121, 122, 123, 124, 125, 126, 142
  job 142
  techniques 121
Seismic 153, 154, 155, 156, 158
  activity 155
  waves 153, 154, 156, 158
Seismographs, portable 155
Seismologists 154, 159
Selective serotonin reuptake inhibitors (SSRIs) 149
Sensor(s) 1, 13, 19, 20, 22, 65, 68, 71, 72, 73, 75, 76, 117, 118, 128
  embedded 75
  fusion 13
  thermal 1
Serotonin production 145
Signal, chemical 139
Single value decomposition (SVD) 223
Sleep disorders 146
Sliding window brute force method 219
SoftMax function 23, 25, 219
Software 93, 129
  robotic process automation 93
  technology 93
Solar 34, 35, 41, 54, 190, 192
  energy 35
  frames 34
  radiation 54
Storage method 105
Support vector machines (SVMs) 90, 91, 92, 214, 215, 220, 221
Swap-aware garbage collection (SAGC) 176, 178

T

Tata consultancy services (TCS) 141, 147
Techniques 13, 15, 26, 43, 45, 46, 47, 51, 52, 66, 124, 177, 181, 216
  contemporary production 66
  intelligent 216
  lightweight cryptography 26
  oscillation 52
Technologies 70, 73, 86, 105
  cloud storage 105
  finance 86
  industrial 73
  sensor 70
Traditional methods 48
Transactions, consumer 93
Transformation, digital 64, 70, 75
Translational bioinformatics 121
Transmission control protocol (TCP) 129
Transportation vehicles 72

V

Virtual prototypes 71
Visual odometry 215

W

Warehouse management 117, 130
Wilcoxon rank sum 56, 209
  method 209
  test 56
Wilcoxon's test 46
Wireless 14, 47, 68, 124, 127, 128, 193
  sensor networks (WSNs) 14, 47, 124, 127, 128, 193
  technologies and sensors 68